HOW TO PASS

STANDARD GRADE
PHYSICS

Campbell White

HODDER
GIBSON
PART OF HACHETTE LIVRE U|

Acknowledgements

The Publishers would like to thank the following for permission to reproduce copyright material:

Photo credits

Page 20 © Denis Scott/Corbis; page 48 © Sovereign, ISM/Science Photo Library; page 55 © AJ Photo/Hop Americain/Science Photo Library; page 59 © Bill Lyons/Alamy; page 126 Jerry Lodriguss/Science Photo Library.

Acknowledgements

Every effort has been made to trace all copyright holders, but if any have been inadvertently overlooked the Publishers will be pleased to make the necessary arrangements at the first opportunity.

Although every effort has been made to ensure that website addresses are correct at time of going to press, Hodder Gibson cannot be held responsible for the content of any website mentioned in this book. It is sometimes possible to find a relocated web page by typing in the address of the home page for a website in the URL window of your browser.

Hachette's policy is to use papers that are natural, renewable and recyclable products and made from wood grown in sustainable forests. The logging and manufacturing processes are expected to conform to the environmental regulations of the country of origin.

Orders: please contact Bookpoint Ltd, 130 Milton Park, Abingdon, Oxon OX14 4SB. Telephone: (44) 01235 827720. Fax: (44) 01235 400454. Lines are open 9.00–5.00, Monday to Saturday, with a 24-hour message answering service. Visit our website at www.hoddereducation.co.uk. Hodder Gibson can be contacted direct on: Tel: 0141 848 1609; Fax: 0141 889 6315; email: hoddergibson@hodder.co.uk

© **Campbell White 2005, 2008**
First published in 2005 by
Hodder Gibson, an imprint of Hodder Education,
Part of Hachette Livre UK,
2a Christie Street
Paisley PA1 1NB

This colour edition published 2008

Impression number	5 4 3 2 1
Year	2012 2011 2010 2009 2008

Cover photo © Photodisc/Getty Images
Artworks by Peters and Zabransky Ltd and Phoenix Photosetting.
Cartoons © Moira Munro 2005, 2008.

Typeset in 10.5 on 14pt Frutiger Light by Phoenix Photosetting, Chatham, Kent
Printed in Italy

A catalogue record for this title is available from the British Library

ISBN-13: 978-0340-973-967

CONTENTS

INTRODUCTION

 ## *About this book and revision*

This revision guide will:

◆ summarise the Standard Grade Physics course

◆ show you how to answer questions

◆ give you hints to get you through the exam

◆ let you try questions similar to those in the exam

◆ help you organise your notes and your revision so that you can work as effectively as possible

◆ list the basic points of each section of the work

◆ give you confidence for the exam.

This guide is interactive, because revision itself has to be interactive. You are probably wondering how it can possibly be interactive, since there is no mouse, joystick or game controller. But interactive does not necessarily mean to do with computers or games consoles. It means that you have to *use* this book, rather than just glance through it. You have chosen to buy a revision guide rather than another game for your games console and so it makes sense to get as much out of it as you can.

By saying that revision is interactive, I mean that you have to work at it, not just go to your room and look busy with a lot of notes around you! To revise effectively, you need to have a complete set of notes that are organised in the correct order.

Firstly you need to find a place to work – preferably somewhere that you can leave your stuff out at the end of a session without having to clear it away. If you have to tidy up after each session, allow a couple of minutes to gather things up.

Next you have to make sure that you have everything you need:

◆ your Physics notes from school   ◆ paper and a pen or pencil

◆ a calculator   ◆ and most importantly – this book!

Generally your revision will be more effective if you do it in a lot of small sessions rather than one or two mammoth sessions. As soon as you begin to lose concentration, you should consider having a break. But if things are going well, then keep going rather than stopping. Reaching the end of one section of your revision is also a good time to have a break. Only you know when to keep on and when to stop.

This book is written in such a way that you have things to do as you work through it. Each chapter contains a lot of self-assessed questions (SAQs). Suggested answers to these SAQs are given at the end of each section. Throughout the book I have also included a number of exam hints which should make your life easier in the examination. These exam hints cover all the parts of the course, not just the part where they are introduced. The book also includes worked examples for you to read through and a summary of the important points at the end of each section.

The final part of the book comprises a sample Credit examination paper. When you have finished your Physics revision, work through this paper to see if there are any parts of the course about which you are still unclear. Answers are provided to help you monitor your progress.

About the standard grade physics course and exam

The Standard Grade Physics course is divided into seven units. These are:

◆ Telecommunication ◆ Using electricity ◆ Health Physics ◆ Electronics
◆ Transport ◆ Energy matters ◆ Space Physics.

If your notes are on loose-leaf paper and have become disordered, the first stage of your revision is to sort them out into these seven units. This book will help you do this as you work through it.

There are two papers in the Standard Grade Physics exam – General and Credit. Most people sit both, but some only sit the General paper. If you are only sitting the General paper, then you can ignore the work in this book that is marked 'Credit level only'. This is shown by the icon ⓒ. Make sure you know what papers you have been entered for, and that you know the date and time of both of the papers.

The General paper lasts $1\frac{1}{2}$ hours and is worth a total of 80 marks.

Hints and Tips

Ninety minutes for 80 marks means that if you work at a rate of one mark per minute, you will have some time for checking your answers at the end.

Of the 80 marks, half will test your knowledge and understanding of Physics. The other 40 marks will test how well you can solve problems based on this knowledge. Solving problems does not just mean 'doing sums', although this is included. It also tests how well you can select information, comment on how good an experiment is, come to a conclusion about information given and explain different bits of Physics.

The Credit examination-style paper at the end of the book includes a mark scheme to enable you to see which parts of the question are testing knowledge and understanding (KU) and which are testing problem solving (PS). Half of the 80 marks are given for each skill.

The Credit paper lasts $1\frac{3}{4}$ hours and is worth a total of 100 marks. Once again, this means that you have to work at a rate of about one mark per minute. For Physics, you need to do some basic maths, so here is your first calculation.

Question

SAQ 1 If you work at a rate of one mark per minute in the Credit Physics exam, calculate how much time you will have to check over your answers at the end of the exam.

For both the General and the Credit papers you do not have a choice of which questions to do – you must attempt them all.

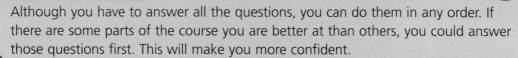

Hints and Tips

Although you have to answer all the questions, you can do them in any order. If there are some parts of the course you are better at than others, you could answer those questions first. This will make you more confident.

If you do answer the questions in a different order to the way they appear, remember to go through the paper at the end and check that you have not missed any out. This particularly applies to the multiple choice questions at the start of the General paper. It is best to know the work thoroughly, but if you really do not know the answer to a multiple choice question, don't leave it blank – have a guess (preferably after you have eliminated the unlikely choices). This way you have at least a one in five chance of being correct. Both the General and the Credit papers cover the work of all seven units of the Standard Grade course.

Some questions in both the General and Credit papers need Physics data, such as the speed of light in glass or the gravitational field strength on Mars. You are not expected to remember these values. You will be given the values in the question paper. For the General paper, all the information needed is given in the question. For the Credit paper, the data is given in a data sheet similar to the one in this book (see page viii). If you have to refer to the data sheet in the exam, the examiner knows that this takes time and so usually gives you a mark for writing the value down in your answer. Don't forget to give the units as well as the value.

Question

SAQ 2 Use the data sheet to find:
a) the speed of light in glass;
b) the gravitational field strength on Mars.

Finally, you will be expected to use various relationships throughout the Standard Grade course. These relationships are given below. However, it is not a lot of good simply memorising these relationships without knowing what they mean and how to use them. As you work through this revision guide, you will become more familiar with them and how to use them.

Answers

Answer to SAQ 1

$1\frac{3}{4}$ hours = 105 minutes
at one mark per minute, 100 marks takes 100 minutes so
time to check over = 105 − 100 = 5 minutes

Answer to SAQ 2

a) The speed of light in glass is $2 \cdot 0 \times 10^8$ m/s

b) The gravitational field strength on Mars is 4 N/kg

Relationships needed for Standard Grade Physics

$s = vt$

$v = f\lambda$

$Q = It$

$E = Pt$

$P = IV$

$P = I^2R$

$P = \dfrac{V^2}{R}$

$V = IR$

$R_T = R_1 + R_2 + \dots$

$\dfrac{1}{R_T} = \dfrac{1}{R_1} + \dfrac{1}{R_2} + \dots$

$P = \dfrac{1}{f}$

$\dfrac{V_1}{V_2} = \dfrac{R_1}{R_2}$

$V_2 = \left(\dfrac{R_2}{R_1 + R_2}\right)V_s$

$V_{gain} = \dfrac{V_o}{V_i}$

$P_{gain} = \dfrac{P_o}{P_i}$

$a = \dfrac{\Delta v}{t}$

$a = \dfrac{v - u}{t}$

$W = mg$

$F = ma$

$E_w = Fs$

$E_p = mgh$

$E_k = \frac{1}{2}mv^2$

$E_h = cm\Delta T$

$E_h = ml$

$\dfrac{n_s}{n_p} = \dfrac{V_s}{V_p} = \dfrac{I_p}{I_s}$

Percentage efficiency $= \dfrac{\text{useful } E_o}{E_i} \times 100$

Percentage efficiency $= \dfrac{\text{useful } P_o}{P_i} \times 100$

Data sheet

Speed of light in materials

Material	Speed in m/s
air	$3\cdot0\times10^8$
carbon dioxide	$3\cdot0\times10^8$
diamond	$1\cdot2\times10^8$
glass	$2\cdot0\times10^8$
glycerol	$2\cdot1\times10^8$
water	$2\cdot3\times10^8$

Speed of sound in materials

Material	Speed in m/s
aluminium	5 200
air	340
bone	4 100
carbon dioxide	270
glycerol	1 900
muscle	1 600
steel	5 200
tissue	1 500
water	1 500

Specific heat capacity of materials

Material	Specific heat capacity in J/kg °C
alcohol	2 350
aluminium	902
copper	386
glass	500
glycerol	2 400
ice	2 100
lead	128
silica	1 033
water	4 180

SI prefixes and multiplication factors

Prefix	Symbol	Factor
giga	G	$1\,000\,000\,000 = 10^9$
mega	M	$1\,000\,000 = 10^6$
kilo	k	$1\,000 = 10^3$
milli	m	$0\cdot001 = 10^{-3}$
micro	μ	$0.000\,001 = 10^{-6}$

Melting and boiling points of materials

Material	Melting point in °C	Boiling point in °C
alcohol	−98	65
aluminium	660	2 470
copper	1 077	2 567
glycerol	18	290
lead	328	1 737
turpentine	−10	156

Specific latent heat of fusion of materials

Material	Specific latent heat of fusion in J/kg
alcohol	$0\cdot99\times10^5$
aluminium	$3\cdot95\times10^5$
carbon dioxide	$1\cdot80\times10^5$
copper	$2\cdot05\times10^5$
glycerol	$1\cdot81\times10^5$
water	$3\cdot34\times10^5$

Specific latent heat of vaporisation of materials

Material	Specific latent heat of vaporisation in J/kg
alcohol	$11\cdot20\times10^5$
carbon dioxide	$3\cdot77\times10^5$
glycerol	$8\cdot30\times10^5$
water	$22\cdot60\times10^5$

Gravitational field strengths

Astronomical object	Gravitational field strength on the surface in N/kg
Earth	10
Jupiter	26
Mars	4
Mercury	4
Moon	1·6
Neptune	12
Saturn	11
Sun	270
Venus	9

Chapter 1

TELECOMMUNICATION

Modern telecommunications systems include telephones (both fixed and mobile), radio, television and computers. These different forms of communication can be interconnected by various methods – by wires, by optical fibre cables or by radio, television or microwave links. Satellites, dish aerials and curved reflectors may be used in the transmission of the signals. This chapter looks at the Physics behind all of this technology used by the telecommunication industry.

Sound and waves

Sound

Sound travels fast, but light travels faster.

You do not need to remember the speed of sound or the speed of light in air (information like this is given in the question for the General paper and on the data sheet in the Credit paper), but you do need to know that sound travels a lot slower than light does.

Question

SAQ 1 Use the data sheet to find the speed of sound in air and the speed of light in air.

You should be able to describe a method of measuring the speed of sound in air. All of the methods use the relationship below.

$$\text{speed} = \frac{\text{distance}}{\text{time}} \text{ which can be written as } v = \frac{d}{t} \text{ or } v = \frac{s}{t}$$

Although far slower than the speed of light in air, the speed of sound in air is still very fast. All the methods of measuring the speed of sound must therefore use either a long distance or a very accurate timer. This is so that your reaction time does not cause you to get a poor result.

Example

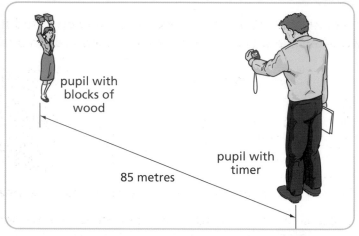

Figure 1.1 Measuring the speed of sound outdoors

pupil with blocks of wood

85 metres

pupil with timer

Two pupils stand 85 metres apart. One pupil hits two blocks of wood together and the other pupil records a time of 0·25 second between seeing the blocks hit and hearing the sound. Calculate the speed of sound that the pupils obtain.

Solution

distance s = 85 metres
time t = 0·25 second
speed v = ?

$$v = \frac{s}{t}$$

$$= \frac{85}{0·25}$$

$$= 340 \text{ metres per second}$$

Hints and Tips

Follow a set of simple rules when you have to do a calculation like the one above.

1 Read the question and list all the quantities given along with the units of each.
2 Convert units if necessary, for example convert centimetres to metres.
3 Include the quantity you have to calculate. In the example above this is speed.
4 Decide on the relationship to use by looking at the quantities you have listed.
5 Carry out the calculation and include the unit in the final answer.

Question

SAQ 2 Why must an electronic timer and not a hand-operated stopwatch be used when the speed of sound is being measured in a laboratory?

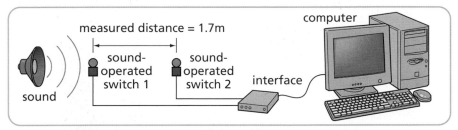

Figure 1.2 Using sound-operated switches to measure the speed of sound

SAQ 3 Two sound-operated switches are connected to a millisecond timer and placed 1·7 metres apart. A sound is made and the timer records the time for the sound to travel between the switches as 0·005 second.

Calculate the speed of sound in air that these results give. (Use the set of rules given on page 2 to help you.)

Waves

Sound is a **wave** that transfers energy, so waves can be used to transmit **signals**.

There are several terms to do with waves that you should be able to use correctly. Some of these are shown in Figure 1.3.

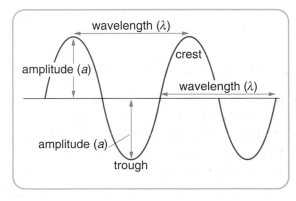

Figure 1.3 Some terms associated with waves

Remember

- The **crest** of a wave is the top of the wave.
- The **trough** of a wave is the bottom of the wave.
- The **wavelength** λ, measured in metres (m), is the distance from any point on a wave to the next corresponding point. Wavelength is the shortest repeating distance.
- The **amplitude** a is the distance, measured in metres (m), from the rest position to the top of a crest or to the bottom of a trough.
- The **frequency** f is the number of waves made in a given time period, usually one second.

$$\text{frequency} = \frac{\text{number of waves}}{\text{time taken}}$$

Frequency is measured in hertz (Hz) where 1 Hz is one wave per second.

- The speed of a wave is the distance travelled in a given time period. So

$$\text{speed} = \frac{\text{distance}}{\text{time}} \qquad v = \frac{s}{t}.$$

Hints and Tips

Do not mix up units in a question. The best way to be sure is to convert all values given into SI units. If distance is given in millimetres, for example, convert this to metres by dividing by 1 000. If time is given in minutes, convert to seconds by multiplying by 60.

Example

A water wave travels a distance of 900 millimetres in a time of 3 seconds. Calculate the wave speed.

Solution

distance s = 900 millimetres

$$= \frac{900}{1\,000} \text{ metre}$$

$$= 0{\cdot}9 \text{ metre}$$

time $t = 3$ seconds

speed $v = ?$

$$v = \frac{s}{t}$$

$$= \frac{0{\cdot}9}{3}$$

$$= 0{\cdot}3 \text{ metres per second}$$

Speed, frequency and wavelength for all waves are linked by the relationship

$$\text{speed} = \text{frequency} \times \text{wavelength} \qquad v = f\lambda$$

Example

A water wave has a frequency of 2 Hz and travels at a speed of 5 m/s.

Calculate the wavelength.

Solution

frequency $f = 2\,\text{Hz}$
speed $v = 5\,\text{m/s}$
wavelength $\lambda = ?$

$$v = f\lambda$$

$$\text{so } \lambda = \frac{v}{f}$$

$$= \frac{5}{2}$$

$$= 2{\cdot}5\,\text{m}$$

Hints and Tips

In the General paper, the units of quantities are written out in words, for example 'metres per second'. In the Credit paper, the units are written as symbols, such as 'm/s'.

You can use any correct form of writing units in your answers.

From now on in this book, I will use symbols (the same as the Credit paper), as in the example above.

What You Should Know

◆ Light travels faster than sound in air.
◆ How to measure the speed of sound in air.
◆ How to do calculations using the formula $\text{speed} = \dfrac{\text{distance}}{\text{time}}$ for sound and water waves.
◆ What the following terms mean: wave, frequency, wavelength, speed, energy, amplitude.
◆ How to do calculations using the formula $\text{speed} = \text{frequency} \times \text{wavelength}$ for sound and water waves.

Answers

SAQ 1

Speed of sound in air = 340 m/s
Speed of light in air = $3{\cdot}0 \times 10^8$ m/s (300 000 000 metres per second)

Answers *continued*

SAQ 2

When the speed of sound is being measured in a laboratory, the time the sound takes to travel is so short that human reaction time is greater than the time being measured.

SAQ 3

distance $s = 1.7$ metres
time $t = 0.005$ second
speed $v = ?$

$$v = \frac{s}{t}$$
$$= \frac{1.7}{0.005}$$
$$= 340 \text{ metres per second}$$

Communicating using cables

The telephone

Messages can be coded as electrical signals and sent along wires. This process is used in the telephone system. The telephone is an example of a long-range communication system. Coded messages are sent out by a **transmitter** and are picked up by a **receiver**.

The mouthpiece of the telephone (the transmitter) contains a microphone that transforms sound energy to electrical energy

The earpiece of the telephone (the receiver) contains an earphone or loudspeaker that transforms electrical energy to sound energy

Electrical signals travel along wires at a speed of almost 300 000 000 m/s

Figure 1.4 How a telephone works

Question

SAQ 4 a) What does the mouthpiece of a telephone contain?

b) What does the earpiece of a telephone contain?

c) State the energy transformation in a microphone.

d) State the energy transformation in a loudspeaker.

e) What is transmitted along wires during a telephone conversation?

The electrical signals in the wires can be displayed as traces on an **oscilloscope** screen. These traces carry information about the sounds that made the electrical signals.

> ## Question
>
> **SAQ 5** Complete the following diagrams to show the signal patterns seen on an oscilloscope screen when the sound signals are as stated.
>
>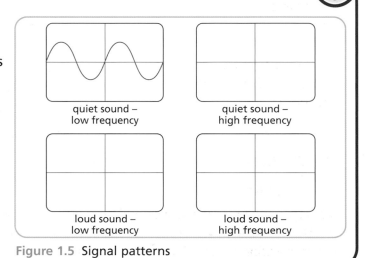
>
> Figure 1.5 **Signal patterns**

Optical fibre communications

Some telecommunication systems use **optical fibres** that carry light signals.

> ## Question   **SAQ 6**   What is an optical fibre?

When a telecommunication system uses optical fibre links, the sound is first transformed into electrical signals using a microphone. These electrical signals are then transformed into light pulses using a **light emitting diode** (LED). These light pulses are transferred along the optical fibre link at very high speed. At the receiver, a **photodiode** transforms the light pulses into electrical signals, which are then changed into sound waves by a loudspeaker.

> ## Question
>
> **SAQ 7** Make a table comparing the properties of electrical wires and optical fibres. Your table should include such factors as: size, cost, weight, signal speed, signal capacity, signal quality, signal reduction.

Example

Light from the Sun takes 8 minutes to reach the Earth. Calculate the distance between the Sun and the Earth.

Solution

time t = 8 minutes = 8×60 = 480 s

speed $v = 3 \times 10^8$ m/s (from data sheet – light travels at the same speed in a vacuum as in air)

distance s = ?

$$v = \frac{s}{t}$$

$$\therefore s = vt$$

$$= 3 \times 10^8 \times 480$$

$$= 1 \cdot 44 \times 10^{11} \text{ m}$$

Hints and Tips

The answer to the example above, $1 \cdot 44 \times 10^{11}$ m, may appear in different ways on the display of your calculator. This is a shorthand way of writing very large numbers. This number is 1·44 followed by 11 zeros. Be careful to write answers like this down in the correct way. If you write the answer down incorrectly as $1 \cdot 44^{11}$ m, you will lose half a mark.

Question

SAQ 8 A glass optical fibre communication link connects two repeater stations that are 50 km apart.

    a) Use the data sheet to find the speed of a signal along the link.

    b) Calculate how long it takes the signal to travel from one station to the other.

Reflection of light

Light can be reflected. This happens when light 'bounces off' a plane (flat) mirror.

Remember

There are two rules to do with the reflection of light.

1   Light reflects off a mirror at the same angle as it meets the mirror. This rule is often stated as 'The angle of incidence is equal to the angle of reflection.'

2   If the direction of a ray of light is reversed, its reflection is in the direction of the original incident ray.

Question

SAQ 9 Add the following labels to the diagram: mirror, incident ray, normal, reflected ray, angle of incidence, angle of reflection.

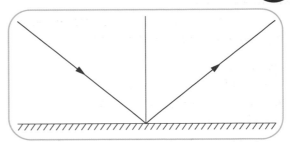

Figure 1.6 The reflection of light from a plane mirror

Sometimes a ray of light reflects inside a glass block. This happens when the angle of incidence is greater than the **critical angle**.

This effect is known as **total internal reflection**. It is because of total internal reflection that light travels along an optical fibre.

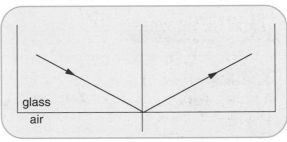

Figure 1.7 Total internal reflection of a ray

Question

SAQ 10 Complete the diagram to show how a signal is transmitted along an optical fibre.

Figure 1.8 An optical fibre

Hints and Tips

The question above is often asked in the exam. It is usually worth two marks and is an easy two marks to get as long as you take care with your drawing. Light travels in straight lines, so you must use a ruler. If you don't, you will lose a mark. You will also lose a mark if you show the light zig-zagging backwards and forwards too many times. The rule that is used is that if you show six or more reflections, you will not gain full marks. Now go back and mark your answer.

HOW TO PASS STANDARD GRADE PHYSICS

What You Should Know

- The purposes of the transmitter and the receiver in a communication system.
- What makes up the mouthpiece and the earpiece of a telephone, and the energy transformations in each.
- About the electrical signals in the wires of a telephone system.
- How the electrical signals in the wires relate to the sounds that produce them.
- What an optical fibre is.
- About the use of optical fibres in telecommunication.
- The differences between electrical cables and optical fibres.
- How to do calculations using the relationship $\text{speed} = \dfrac{\text{distance}}{\text{time}}$ for light.
- About reflection of light from a plane mirror.

Answers

SAQ 4

a) The mouthpiece of a telephone contains a microphone.

b) The earpiece of a telephone contains a loudspeaker (or an earphone).

c) The energy transformation in a microphone is sound to electrical.

d) The energy transformation in a loudspeaker is electrical to sound.

e) Electrical signals are transmitted along wires during a telephone conversation.

SAQ 5

quiet sound –
low frequency

quiet sound –
high frequency

loud sound –
low frequency

loud sound –
high frequency

Figure 1.9 Signal patterns

SAQ 6

An optical fibre is a very thin, high quality strand of glass that can carry light signals at very high speed.

Answers continued

SAQ 7

	Electrical wires	Optical fibres
size	large diameter	small diameter
cost	expensive	cheap
weight	wire is heavy	fibres are not so heavy
signal speed	travel faster	travel slower (but still very fast)
signal capacity	fewer signals per wire	more signals per fibre
signal quality	electrical interference and can be 'tapped into' easily	no electrical interference and difficult to 'tap into'
signal reduction	repeaters needed every 4 km to boost signal (more energy loss per kilometre)	repeaters needed only every 100 km to boost signal (less energy loss per kilometre)

SAQ 8

a) Speed of light in glass = $2 \times 10^8 \, \text{m/s}$

b) Distance $s = 50 \, \text{km} = 50 \times 10^3 \, \text{m}$
   speed $v = 2 \times 10^8 \, \text{m/s}$
   time $t = ?$

$$v = \frac{s}{t}$$

$$t = \frac{s}{v}$$

$$= \frac{50 \times 10^3}{2 \times 10^8}$$

$$= 2 \cdot 5 \times 10^{-4} \, \text{s}$$

SAQ 9

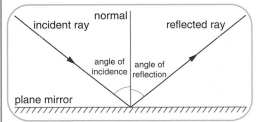

Figure 1.10 Reflection of light

SAQ 10

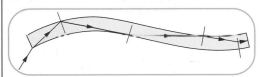

Figure 1.11 Light transmission in an optical fibre

Radio and television

Radio waves

There are three forms of long-range communication that do not need cables between the transmitter and the receiver – mobile phones, radio and television.

> ### Remember
>
> Here are some facts about the microwaves and the television and radio signals that are sent during long range communication:
>
> ◆ They are all waves.
> ◆ They all transfer energy.
> ◆ They are transmitted at very high speed.
> ◆ They are transmitted through air at 300 000 000 m/s.

Example

C A student in Glasgow speaks on her mobile phone to her friend in London 660 km away. The mobile phones use a microwave link.

Calculate the time it takes the microwave signal to travel from Glasgow to London.

Solution

Distance $s = 660\,\text{km} = 660{\times}10^3\,\text{m}$

speed $v = 300\,000\,000\,\text{m/s} = 3{\times}10^8\,\text{m/s}$

time $t = ?$

$$v = \frac{s}{t}$$

$$\therefore t = \frac{s}{v}$$

$$= \frac{660{\times}10^3}{3{\times}10^8}$$

$$= 0{\cdot}0022\,\text{s}$$

Radio and television receivers

Consider a radio receiver as being made up of six separate parts. You will need to know these parts, be able to identify them on a block diagram and know what their purpose is. They are: the aerial, the tuner, the decoder, the amplifier, the loudspeaker and the electricity supply.

Question

SAQ 11 Using the names of the parts given on page 12, label each of the blocks on the block diagram of a radio receiver shown below.

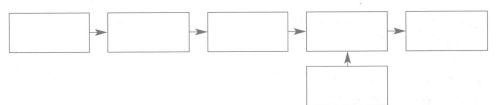

Figure 1.12

SAQ 12 Match the parts of a radio receiver in the left-hand column with the functions in the right-hand column by writing out the complete sentences.

The aerial	boosts or increases the energy of the audio signal.
The tuner	converts the electrical audio signals into sound waves that can be heard.
The decoder	detects radio waves from all transmitters in range and changes these into high-frequency electrical signals.
The amplifier	picks out the electrical signals from the radio station that you want to listen to.
The loudspeaker	provides the extra energy needed for the audio signals to be amplified.
The electricity supply	separates out the audio part of the high-frequency electrical signals.

Figure 1.13

Hints and Tips

A good way to learn something is to write it out yourself. By writing it, you are forced to think more about the work. I could have asked you to answer the question above by simply drawing lines between the parts on the left and the functions on the right, but this would not have been as good a way to learn the functions.

A television receiver can also be thought of as several parts, each with its own function. Although it contains more parts than in a radio receiver, the block diagram of a television receiver is similar. The main parts are: the aerial, the tuner, the audio decoder, the video decoder, the audio amplifier, the video amplifier, the picture tube, the loudspeaker and the electricity supply.

Question

SAQ 13 Using the names of the parts given above, label each of the blocks on the block diagram of a television receiver shown below.

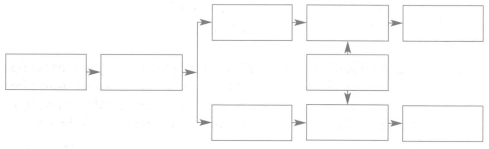

Figure 1.14

SAQ 14 Match the parts of a television receiver in the left-hand column with the functions in the right-hand column by writing out the complete sentences.

The aerial	boosts or increases the energy of the audio part of the signal.
The tuner	boosts or increases the energy of the video (picture) part of the signal.
The audio decoder	converts the electrical audio signals into sound waves that can be heard.
The video decoder	converts the video part of the electrical signal into a picture made up of light on the screen.
The audio amplifier	detects television waves from all transmitters in range and changes these into high-frequency electrical signals.
The video amplifier	picks out the electrical signals from the television station that you want to watch and listen to.
The loudspeaker	provides the extra energy needed for the audio and video signals to be amplified.
The picture tube	separates out the audio part of the high-frequency electrical signals.
The electricity supply	separates out the video (picture) part of the high-frequency electrical signals.

Figure 1.15

Television pictures

A picture is produced on a television screen by a beam of electrons striking the phosphor-coated surface of the screen. When the electrons strike the phosphor, the screen emits light. The more electrons that hit a point on the screen, the brighter is that point.

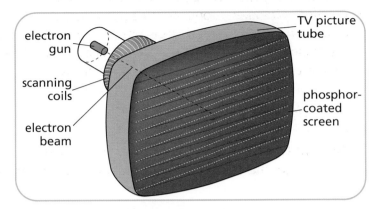

Figure 1.16 A TV picture tube

Electronics inside the television make the beam of electrons move from left to right and from top to bottom of the screen, just like your eyes do when you read a book. This process is called **scanning**. In this way a picture is built up as the lines are traced on the screen. Figure 1.16 shows how a picture is produced on a television screen.

Question

SAQ 15 Describe how a *moving* picture is seen on a television screen. Include the ideas 'line build-up', 'image retention' and 'brightness variation' in your description.

(This is quite a hard question, even at Credit level, but make an attempt before you look at the answer given.)

Mixing red, green and blue lights produces all the colours seen on a colour television screen. There are three electron guns in a colour picture tube. Each gun aims an electron beam through holes in a shadow mask behind the screen. The phosphor coating consists of a large number of dots in sets

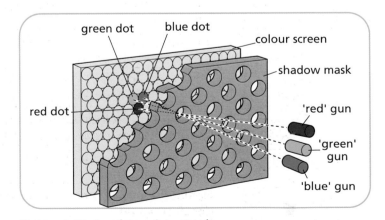

Figure 1.17 A colour picture tube

of three. Each dot gives out red, green or blue light when hit. The shadow mask makes sure that each electron beam hits only the correct colour of dots.

HOW TO PASS STANDARD GRADE PHYSICS

Hints and Tips

Make sure that you write exactly what you mean to when you give a description. It is important to be precise with the language you use in Physics. A good example is the description of how a colour picture tube works. Do not write about 'red, green and blue electrons'. Do not confuse the electron beam with a beam of light. The examiner cannot give you any marks for this. Even though you know what you meant to write, the examiner can only give you marks for correctly written Physics.

C White light is produced when red, green and blue lights are mixed. Other colours are produced when pairs of the lights are mixed.

Question

C **SAQ 16** Complete the following diagram to show the colours produced in the overlapping sections when coloured lights are mixed.

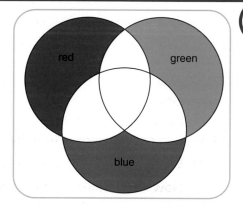

Figure 1.18 Colour mixing

Transmission and reception

If you are only doing General Level work, all you need to know about transmission and reception of radio waves is that a radio transmitter can be identified by the wavelength or the frequency of the carrier wave it uses.

Question

SAQ 17 Use a radio listings magazine to find the wavelength of one radio station and the frequency of a different radio station.

C The rest of the work on transmission and reception is Credit Level only.

The main stages in radio transmission are shown in Figure 1.19.

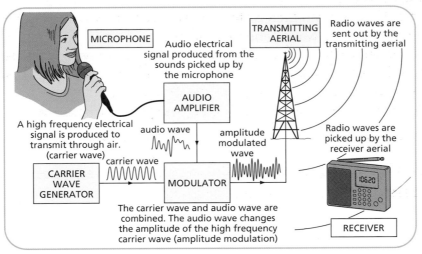

Figure 1.19 Radio transmission

Use the diagram to answer the question below.

Question ?

C **SAQ 18 a)** In a radio transmitter, why is the carrier wave generator needed?

b) What is meant by amplitude modulation?

c) How does information get from a transmitter to a receiver?

C The general principle of television transmission is similar to radio transmission with some differences.

◆ The carrier wave generated is **ultra high frequency** (UHF) so that more information can be transmitted.

◆ The carrier wave is **modulated** by a video signal as well as an audio signal.

◆ The carrier wave has its frequency modulated by the video and audio signals. This is known as **frequency modulation** (FM). Frequency modulation is also used for **very high frequency** (VHF) radio transmissions.

◆ The modulated radio wave is transmitted and picked up by the aerial of a television receiver. The wave is **demodulated** and the video and audio signals separated.

Radio waves used for communication are grouped into different bands depending on their properties and uses.

Question

SAQ 19 Use the information given in the table to answer the questions about radio wave bands.

Frequency name	Frequency range	Wavelength name	Property/uses
extra low frequency (ELF)	30 Hz–3 kHz		the only waves that travel far in water
low frequency (LF)	30 kHz–300 kHz	long wave	long range communication
	300 kHz–3 MHz	medium wave	sound broadcasts
very high frequency (VHF)	30 MHz–300 MHz		high-quality sound broadcasts travel in straight lines
ultra high frequency (UHF)	300 MHz–3 000 MHz		line of sight communication only television broadcasts
	greater than 3 000 MHz	microwaves	the only waves that pass through the ionosphere round the Earth

a) Explain why ELF waves are used to communicate with submarines which are moving in deep oceans.

b) Which wave would be used for a high-quality sound broadcast – one with a frequency of 2 MHz or one with a frequency of 200 MHz?

c) Explain why microwaves are used for satellite communication.

When waves pass the edge of an obstacle, they **diffract** or bend round the edge. Longer wavelength, lower frequency waves diffract more than shorter wavelength, higher frequency waves. Because of diffraction, obstacles such as hills can affect radio and television reception.

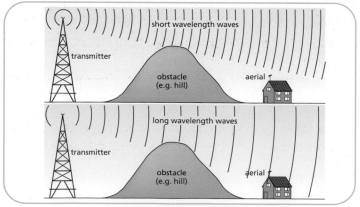

Figure 1.20 The diffraction of waves by obstacles

Question

SAQ 20 Use Figure 1.20 to explain whether it is radio reception or television reception that is poorer in hilly areas.

You will need to be able to do calculations using the relationship speed = frequency × wavelength for microwaves, television and radio waves.

Example

Radio 5 Live broadcasts on 909 kHz. Calculate the wavelength of the transmitted wave.

Solution

Speed of radio waves $v = 3\times10^8\,\mathrm{m/s}$
frequency $f = 909\,\mathrm{kHz} = 909\times10^3\,\mathrm{Hz}$
wavelength $\lambda = ?$

$$v = f\lambda$$
$$\text{so } \lambda = \frac{v}{f}$$
$$= \frac{3\times10^8}{909\times10^3}$$
$$= 330\cdot033$$
$$\text{so wavelength} = 330\,\mathrm{m}$$

Hints and Tips

The number shown on the calculator as the answer to the problem above is 330·033. However you must state your answer to a sensible number of significant figures. It is best to give your answer to the same number of significant figures as the numbers in the question, but up to two figures more or one figure less is acceptable.

In the example above, the frequency was given to three significant figures, so I have given the answer to three significant figures.

Satellites and dish aerials

There are many satellites orbiting the Earth. The time it takes a satellite to make one revolution around the Earth (its **period**) depends on the height of the satellite above the Earth.

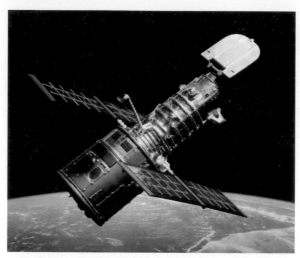

A satellite in orbit

Satellites that orbit such that they always stay above the same point on the Earth's surface are called **geostationary satellites**.

Question **SAQ 21** What is the period of a geostationary satellite?

Curved reflectors that form part of dish aerials or receivers increase the strength of the signal that is received. They do this by collecting a lot of the television waves or microwaves that would otherwise pass by the detector.

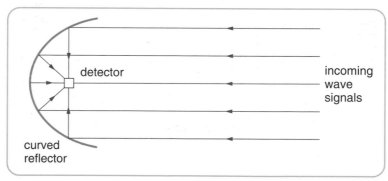

detector

incoming wave signals

curved reflector

Figure 1.21 A curved reflector and detector

Curved reflectors are also used with certain transmitters of television waves and microwaves. The source of the waves to be transmitted is placed at the focus of a curved reflector. In this way, a parallel-sided beam of waves can be sent to a receiving aerial. Because of this, more of the energy can be transmitted to the receiving aerial.

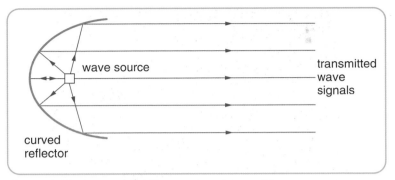

Figure 1.22 A curved reflector on a transmitter

Satellites and curved reflectors are used in telecommunication, in particular intercontinental telecommunication. You should be able to describe an application of curved reflectors, the principle of satellite television and the principle of intercontinental telecommunication. To do this you need to know the purpose of boosters, repeaters, ground stations and geostationary satellites. Figure 1.23 shows how a signal, such as a telephone call, could be transmitted and received.

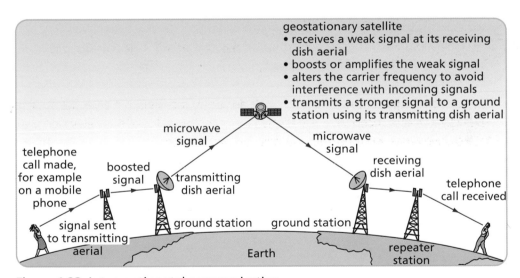

Figure 1.23 Intercontinental communication

Question

SAQ 22 a) What type of signal is sent to a geostationary satellite?

b) What are the four things that happen to the signal at a geostationary satellite?

HOW TO PASS STANDARD GRADE PHYSICS

What You Should Know

C
- ◆ About microwaves, television and radio signals.
- ◆ How to do calculations using the relationship $\text{speed} = \dfrac{\text{distance}}{\text{time}}$ for microwaves, television waves and radio waves.
- ◆ About the main parts of a radio receiver.
- ◆ About the main parts of a television receiver.

C
- ◆ How a picture is produced on a television screen and, for Credit Level only, how a moving picture is seen on a television screen.
- ◆ About colour television pictures.

C
- ◆ The general principles of radio and television transmission.
- ◆ Some properties and uses of radio waves and radio bands.
- ◆ How to do calculations using the relationship $\text{speed} = \text{frequency} \times \text{wavelength}$ for microwaves, television waves and radio waves.

C
- ◆ About curved reflectors on certain receivers and, for Credit Level only, on certain reflectors.
- ◆ What the period of a satellite's orbit depends on.
- ◆ What is meant by a geostationary satellite.
- ◆ How geostationary satellites, ground stations and dish aerials are used for intercontinental telecommunication.

Answers

SAQ 11

| aerial | → | tuner | → | decoder | → | amplifier | → | loudspeaker |

electricity supply → amplifier

Figure 1.24

SAQ 12

The **aerial** detects radio waves from all transmitters in range and changes these into high-frequency electrical signals.

The **tuner** picks out the electrical signals from the radio station that you want to listen to.

The **decoder** separates out the audio part of the high-frequency electrical signals.

Answers *continued*

The **amplifier** boosts or increases the energy of the audio signal.

The **loudspeaker** converts the electrical audio signals into sound waves that can be heard.

The **electricity supply** provides the extra energy needed for the audio signals to be amplified.

SAQ 13

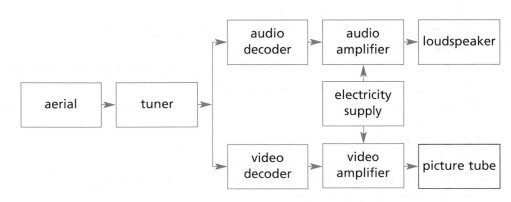

Figure 1.25

SAQ 14

The **aerial** detects television waves from all transmitters in range and changes these into high-frequency electrical signals.

The **tuner** picks out the electrical signals from the television station that you want to watch and listen to.

The **audio decoder** separates out the audio part of the high-frequency electrical signals.

The **video decoder** separates out the video (picture) part of the high-frequency electrical signals.

The **audio amplifier** boosts or increases the energy of the audio part of the signal.

The **video amplifier** boosts or increases the energy of the video (picture) part of the signal.

The **loudspeaker** converts the electrical audio signals into sound waves that can be heard.

The **picture tube** converts the video part of the electrical signal into a picture made up of light on the screen.

The **electricity supply** provides the extra energy needed for the audio and video signals to be amplified.

Answers *continued*

SAQ 15

A moving picture on a television screen is a series of rapidly changing still pictures. Each still picture is built up as a series of lines by an electron beam scanning the surface of the screen.

As the electron beam moves across the screen, the video signal changes the number of electrons in the beam that hit the screen at different places. This causes brightness variation of the light that is emitted by the phosphor coating. So a picture is built up using varying brightness from white through greys to black.

Even though the electron beam only hits one point of the screen each instant, the phosphor coating continues to glow for long enough for our eyes to see a complete picture. Each still picture is then rapidly replaced by the next one, which shows the movement as it is a short time later. Our eyes merge the lines into pictures and the still pictures into moving pictures because of an effect called 'image retention'.

SAQ 16

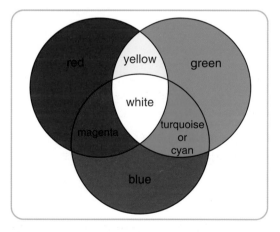

Figure 1.26

SAQ 17

Some wavelengths and frequencies you may have found include the following. (There are lots of others.)

Radio station	Wavelength	Frequency
Radio Scotland	370 m	810 kHz 92·4–94·7 MHz
Radio 1		98·0–99·5 MHz
Radio 4	1 515 m	198 kHz
Radio 5 Live		693 kHz 909 kHz

Answers *continued*

SAQ 18

a) The carrier wave generator is needed to produce the high-frequency electrical signals to transmit through air.

b) Amplitude modulation is the process of combining an audio wave with a carrier wave in such a way that the audio wave changes the amplitude of the carrier wave.

c) Radio waves are sent out by the transmitting aerial and are picked up by the receiver aerial.

SAQ 19

a) Extra low frequency waves are the only waves that travel far in water.

b) A 200 MHz wave would be used for a high-quality sound broadcast.

c) Microwaves are the only waves that pass through the ionosphere round the Earth.

SAQ 20

Television broadcasts use higher frequency, shorter wavelength waves. Shorter wavelengths diffract less, so television reception is poorer in hilly areas.

SAQ 21

The period of a geostationary satellite is 24 hours.

SAQ 22

a) A microwave signal is sent to a geostationary satellite.

b) The geostationary satellite:
   ◆ receives weak signals from a ground station at its receiving dish aerial
   ◆ amplifies the weak signal
   ◆ alters the carrier frequency (to avoid interference)
   ◆ transmits a stronger signal to a ground station using its transmitting dish aerial.

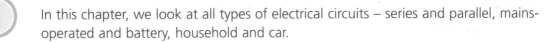

Chapter 2

USING ELECTRICITY

In this chapter, we look at all types of electrical circuits – series and parallel, mains-operated and battery, household and car.

Electrical appliances transfer electrical energy at different rates. The rate at which energy is transferred is the **power** of the appliance. We look at power and energy of appliances. Electricity can be dangerous if it is not treated carefully. Safety in using electricity is an important part of this unit. Some electrical appliances such as the electric motor make use of the magnetic effect of a current. The chapter ends by looking at this effect.

Appliances and supplies

Batteries and the mains supply

All electric circuits have a source of energy – either a battery or the mains supply. Both supply the electrical energy to make the charges move around the circuit. With a battery, the charges always move the same way around a circuit. With the mains supply, the direction of movement of the charges constantly changes.

Question

SAQ 1   a)   Electric supplies are described as a.c. or d.c. What do the terms a.c. and d.c. mean?

            b)   Is the mains an a.c. or a d.c. supply?

            c)   Is a battery an a.c. or a d.c. supply?

            d)   What is the frequency of the mains supply?

            e)   What is the value of the mains voltage?

Hints and Tips

Hopefully you answered 230 volts to part e) of SAQ 1. The value of the mains voltage (also called the 'declared value') is sometimes given as 240 volts in older textbooks. If you are asked the value of the mains voltage in the exam, or have to use the value in a problem, you will lose a mark if you use the wrong value.

Appliances and cables

Household appliances transfer the electrical energy and produce some other type of energy, often heat, light, sound or movement.

Question

SAQ 2 Name appliances that produce each of the four types of energy mentioned above.

The rate at which electrical appliances transfer energy is called the power rating of the appliance. Power is measured in **watts** (W). As a general rule, appliances that are designed to produce heat have higher power ratings.

Question

SAQ 3 Match the power ratings given to the appliances in the following list.

Power ratings: 1 watt, 60 watts, 800 watts, 2 500 watts

Appliance	Power rating
bedside lamp	
hairdryer	
electric clock	
electric kettle	

Appliances are connected to the mains supply through a flex. Usually flexes have three conductors called the **live**, the **neutral** and the **earth** wires. The insulation round each of these wires is colour-coded so that they can be identified.

Question

SAQ 4 Give the name of the wire, and the colour of insulation of each, that should be connected to pins 1, 2 and 3 of the three-pin plug shown.

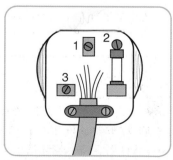

Figure 2.1

HOW TO PASS STANDARD GRADE PHYSICS

Safety with electricity

Most electrical appliances need a three-core flex. But not all.

Question

SAQ 5 How would you recognise an electrical appliance that only needs a two-core flex and which wire is not needed?

An earth wire is a safety device. It acts, along with the fuse in the plug, to stop you from getting an electric shock. (The human body is a conductor of electricity.) It is not so easy to get an electric shock from a **double insulated appliance** because there are two insulating layers to protect you.

Question

SAQ 6 Copy out the following passage about an electric heater, choosing the correct words.

The earth wire is connected to the (heater/metal casing) of the appliance. If a fault happens in the appliance and the live wire makes contact with the casing, the casing would become (live/neutral) If present, the earth wire gives a (low/high) resistance path for the current. This (larger/smaller) current is enough to (operate the heater/blow the fuse). The faulty appliance is no longer connected to the live supply and is not dangerous. If the earth wire is not connected, the casing of the faulty appliance would remain (live/neutral) and therefore (safe/dangerous) to touch. Fuses and switches must be in the (live/neutral) wire. If a switch was in the (live/neutral) wire, the appliance would still be connected to the live supply, even when the switch was (closed/open). If a fuse was in the (live/neutral) wire, the appliance would still be connected to the live supply, even if the fuse had blown.

Some situations to do with electricity could cause accidents (see Figure 2.2).

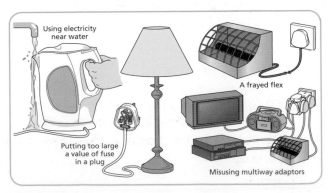

Using electricity near water

A frayed flex

Putting too large a value of fuse in a plug

Misusing multiway adaptors

Figure 2.2

Question

SAQ 7 Link each of the statements in the first column to a statement in the second column by using the phrase '... could be dangerous because ...' and write out the complete sentences.

a) Using electricity near water...

...the current would be larger than it should be, causing overheating.

b) Putting too large a value of fuse in a plug...

...a fault could cause a large current in the flex which could cause overheating.

c) A frayed or badly connected flex...

...a person could touch a live wire and be electrocuted.

d) Allowing a short circuit...

...too much current could be taken from the socket causing the house wiring to overheat.

e) Misusing multiway adaptors...

...water is a conductor and moisture makes the human body a better conductor.

What You Should Know

- The purpose of batteries and the mains supply.
- About the terms a.c. and d.c.
- About the power ratings of household appliances.
- About flexes and fuses for household appliances.
- That the earth wire is a safety device and, for Credit Level only, how it acts as a safety device.
- About double insulation.
- Situations involving electricity that could result in accidents.

Answers

SAQ 1

a) a.c. means 'alternating current'. With an a.c. supply, the charges constantly change direction – they alternate in direction, hence the name. d.c. means 'direct current'. With a d.c. supply, the charges always move in the same direction in the electric circuit.

b) The mains is an a.c. supply.

Answers *continued*

c) A battery is a d.c. supply.

d) The frequency of the mains supply is 50 hertz. This means that the charges go through one complete change of direction 50 times every second.

e) The mains voltage is 230 volts.

SAQ 2

There are many answers you could have given for each, such as: heat – toaster, light – bedside lamp, sound – radio, movement – food mixer.

SAQ 3

Appliance	Power rating
bedside lamp	60 watts
hairdryer	800 watts
electric clock	1 watt
electric kettle	2 500 watts

SAQ 4

Pin 1 – earth wire, green and yellow stripes

Pin 2 – live wire, brown

Pin 3 – neutral wire, blue

SAQ 5

Electrical appliances that have the double insulation symbol do not need an earth wire.

Figure 2.3 **The double insulation symbol**

SAQ 6

The earth wire is connected to the **metal casing** of the appliance. If a fault happens in the appliance and the live wire makes contact with the casing, the casing would become **live**.

If present, the earth wire gives a **low** resistance path for the current. This **larger** current is enough to **blow the fuse**. The faulty appliance is no longer connected to the live supply and is not dangerous.

If the earth wire is not connected, the casing of the faulty appliance would remain **live** and therefore be **dangerous** to touch.

Fuses and switches must be in the **live** wire. If a switch was in the **neutral** wire, the appliance would still be connected to the live supply, even when the switch was **open**. If a fuse was in the **neutral** wire, the appliance would still be connected to the live supply, even if the fuse had blown.

SAQ 7

a) Using electricity near water could be dangerous because water is a conductor and moisture makes the human body a better conductor.

b) Putting too large a value of fuse in a plug could be dangerous because a fault could cause a large current in the flex which could cause overheating.

c) A frayed or badly connected flex could be dangerous because a person could touch a live wire and be electrocuted.

d) Allowing a short circuit could be dangerous because the current would be larger than it should be, causing overheating.

e) Misusing multiway adaptors could be dangerous because too much current could be taken from the socket causing the house wiring to overheat.

Six electrical quantities

Question

SAQ 8   This section is rather mysteriously titled 'Six electrical quantities'. Do you know what these quantities are?

Don't worry if you can't answer this question just now – it's a bit like asking you to predict the conclusion of a mystery story. You will get the chance to answer it again at the end of this section.

Electric circuits

All electric circuits contain **components** or parts that transfer electrical energy as other forms. They also contain a source of electrical energy and, usually, connecting wires. You should know the following circuit symbols: battery, fuse, lamp, switch, resistor, variable resistor, capacitor, diode, voltmeter, ammeter.

Question

SAQ 9   Copy and complete the table of circuit symbols.

Component	Circuit symbol	Component	Circuit symbol
ammeter			—⊗—
	—|·|—	resistor	
capacitor			—/—
	—▷|—	variable resistor	
fuse			—Ⓥ—

Figure 2.4 Circuit symbols for some common components

An electric circuit must be complete with no breaks in it (an **open circuit**) and must have no parts where the ends of a component are connected together by a conductor (a **short circuit**).

Question

SAQ 10 a) Draw the circuit diagram of a continuity tester that contains a battery and a lamp.

b) Explain how the tester is used to check whether a fuse has 'blown'.

When there is a complete electric circuit with a voltage supply, electric charge flows round the circuit. In circuits made up of solid conductors, the charges are carried on electrons that are free to move. The rate at which the charges move around the circuit is called the **current**. Current is measured in **amperes** or **amps** (A) or **milliamps** (mA).

Remember

C For Credit Level only, you need to know the following:

◆ Electric charge is measured in **coulombs** (C).
◆ Charge, current and time are connected by the relationship
charge (in C) = current (in A) × time (in s) or $Q = I\,t$
◆ The supply in an electric circuit is a source of energy. When charges move around a circuit they transfer the energy of the supply to the other components. The voltage of the supply is a measure of the energy transferred by the charges.

Example

C In one minute, 180 coulombs of charge are moved around a circuit. Calculate the current in the circuit.

Solution

charge $Q = 180$ C
time $t = 1$ minute $= 60$ s
current $I = ?$

$$Q = It$$
$$\therefore I = \frac{Q}{t}$$
$$= \frac{180}{60}$$
$$= 3\,\text{A}$$

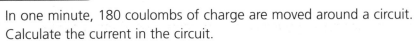

Resistance

All electrical appliances or components oppose the flow of charge – they oppose current. The opposition to current is called **resistance**. The bigger the resistance in a circuit, the smaller the current. Resistance is measured in **ohms** (Ω). The relationship linking resistance, voltage and current is: $\text{resistance} = \dfrac{\text{voltage}}{\text{current}}$ or $R = \dfrac{V}{I}$.

Example

A torch bulb has a voltage of 3 V across it. The current in the bulb is 0·25 A.

Calculate the resistance of the bulb.

Solution

voltage $V = 3$ V
current $I = 0·25$ A
resistance $R = ?$

$$R = \frac{V}{I} \ (\tfrac{1}{2} \text{ mark})$$

$$= \frac{3}{0·25} \ (\tfrac{1}{2} \text{ mark})$$

$$= 12\,\Omega \ (1 \text{ mark})$$

Hints and Tips

You will have noticed that marks have been indicated in the solution to the example above. This type of calculation is what the examiner calls a 'standard two-marker'. With this type of question, the marks are awarded as follows:

- ◆ $\tfrac{1}{2}$ mark for correctly stating the relationship in words or symbols
- ◆ $\tfrac{1}{2}$ mark for correctly substituting the values into the relationship
- ◆ 1 mark for the correct final answer with the correct unit.

If you do not get the correct answer, you could still gain some marks as long as you show all the steps in the answer.

The ratio $\dfrac{V}{I}$ for a resistor stays approximately the same even if the current is changed. (This is only true if the temperature is constant.) Some resistors are designed to have a resistance that can be varied. These are called **variable resistors**.

Question SAQ 11 Give two practical uses for variable resistors.

Energy and power

Whenever there is an electric current in a wire, there is an energy transfer. Quite often when energy is transferred, heat is produced. This is because of the resistance of the electric circuit. Energy is measured in **joules** (J).

Question

SAQ 12 Name three appliances used in the home in which electrical energy is transferred and heat is produced.

The rate at which energy is transferred is called power.

$$\text{power} = \text{rate of transfer of energy}$$
$$= \frac{\text{energy}}{\text{time}}$$
$$P = \frac{E}{t}$$

Example

Calculate the energy transferred in a 100 W light bulb in 5 minutes.

Solution

power $P = 100\,\text{W}$
time $t = 5$ minutes $= 5 \times 60\,\text{s}$
energy $E = ?$

$$P = \frac{E}{t}$$
$$\therefore E = Pt$$
$$= 100 \times 5 \times 60$$
$$= 30\,000\,\text{J}$$

Question

SAQ 13 The joule is not the only energy unit you should know about.

    a) What is the unit of energy that is shown on a domestic electricity meter?

    b) What is the relationship between this unit and the joule?

In any electrical circuit, rate of electrical energy transfer = voltage × current

$$\text{power} = \text{voltage} \times \text{current}$$
$$P = VI$$

Example

A torch bulb has a voltage of 3 V across it. The current in the bulb is 0·25 A. Calculate the power of the bulb.

Solution

voltage $V = 3$ V

current $I = 0.25$ A

power $P = ?$

$P = VI$

$= 3 \times 0.25$

$= 0.75$ W

Question

SAQ 14 Use the relationships $V = I R$ and $P = V I$ to show that there are other power relationships: $P = I^2R$ and $P = V^2/R$.

Example

A torch bulb has a resistance of 12 Ω. The current in the bulb is 0·25 A. Calculate the power of the bulb.

Solution

resistance $R = 12\,\Omega$

current $I = 0.25$ A

power $P = ?$

$P = I^2R$

$= 0.25^2 \times 12$

$= 0.75$ W

Lamps and heaters

Lamps and heaters both transfer energy. Figure 2.5 gives information about the energy transfer in two types of lamp (a filament lamp and a gas discharge tube) and in an electric heater.

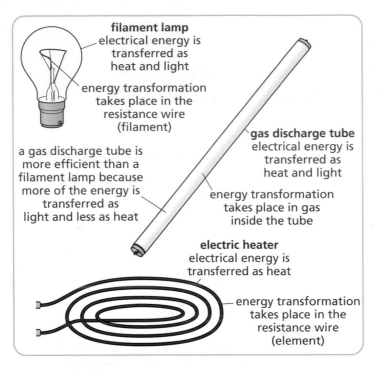

filament lamp
electrical energy is transferred as heat and light

energy transformation takes place in the resistance wire (filament)

a gas discharge tube is more efficient than a filament lamp because more of the energy is transferred as light and less as heat

gas discharge tube
electrical energy is transferred as heat and light

energy transformation takes place in gas inside the tube

electric heater
electrical energy is transferred as heat

energy transformation takes place in the resistance wire (element)

Figure 2.5 Energy transfer in a filament lamp, gas discharge tube and an electric heater

Question

SAQ 15 a) What energy transfer takes place in a lamp?

b) What energy transfer takes place in a heater?

c) Where does the energy transfer take place in a filament lamp?

d) Where does the energy transfer take place in a gas discharge tube?

e) Where does the energy transfer take place in an electric heater?

f) Why is a gas discharge tube more efficient than a filament lamp?

Now that you have finished this section, go back and check your answer to SAQ 8. If you did not manage to answer it before, you should be able to answer it now.

What You Should Know

◆ How to draw and identify various circuit symbols.
◆ What an electric current is.
◆ How to do calculations using the relationship charge = current × time.
◆ What is meant by the voltage of a supply.

What you should know continued ➤

What You Should Know *continued*

◆ What is meant by resistance and how it is measured.

◆ How to do calculations using the relationship $resistance = \dfrac{voltage}{current}$.

◆ About energy transfer in electrical circuits.

◆ The relationship between energy and power.

◆ How to do calculations using the relationship power = voltage × current.

C ◆ How to do calculations using the relationship power = (current)^2 × resistance.

◆ About energy transfer in electric lamps and heaters.

◆ About the efficiency of different types of electric lamps.

Answers

SAQ 8

The six electrical quantities are:

◆ charge, measured in coulombs (C)
◆ current, measured in amperes (A)
◆ voltage, measured in volts (V)

◆ resistance, measured in ohms (Ω)
◆ energy, measured in joules (J)
◆ power, measured in watts (W).

SAQ 9

Component	Circuit symbol	Component	Circuit symbol
ammeter	—(A)—	lamp	—⊗—
battery	—┤├·├—	resistor	—▭—
capacitor	—┤├—	switch	—/ _
diode	—▷|—	variable resistor	—▱—
fuse	—▭—	voltmeter	—(V)—

SAQ 10

a)

Figure 2.6 A continuity tester

b) The ends of the tester are first joined to check that the tester is working (the lamp lights). The ends are then touched to the ends of the fuse being tested. If the lamp lights, the fuse has not blown. If the lamp does not light, the fuse has blown.

Answers continued

SAQ 11

There are many uses, such as: volume control for a radio, brightness control for a television, joystick for a computer game, speed control for an electric motor.

SAQ 12

There are a lot of appliances you could have named, including: an electric fire, a toaster, an electric kettle, an electric cooker, an electric iron.

SAQ 13

a) The unit of energy that is shown on a domestic electricity meter is the kilowatt-hour (kWh).

b) One kilowatt-hour is the energy transferred by a 1 kW (1 000 W) appliance in 1 hour, so:
    1 kilowatt-hour = $1\,000 \times 60 \times 60$ watt-seconds (joules) = $3\,600\,000$ J.

SAQ 14

$P = V I$

so $P = I R \times I$ (since $V = I R$)

so $P = I^2 R$

$P = V I$

so $P = V \times V/R$ (since $I = V/R$)

so $P = V^2/R$

SAQ 15

a) In a lamp, electrical energy is transferred as heat and light.

b) In a heater, electrical energy is transferred as heat.

c) The energy transfer takes place in resistance wire (the filament) in a filament lamp.

d) The energy transfer takes place in the gas inside the discharge tube.

e) The energy transfer takes place in resistance wire (the element) in an electric heater.

f) A discharge tube is more efficient than a filament lamp because more of the energy is transformed into light and less into heat.

More circuits

Series and parallel circuits

You should be able to recognise **series** and **parallel circuits** and know and be able to use some facts about the current and voltage in them.

Remember

◆ In a series circuit, the components come one after the other, like the episodes in a television serial.
◆ A parallel circuit must have at least two junctions of conductors.
◆ Current is a flow of charge around a circuit, so when a moving charge comes to a junction, it can go one of two ways.
◆ Voltage appears across two points in a circuit.

Questions

SAQ 16 Copy and complete the statements about current and voltage in series and parallel circuits by using the phrases 'is the same as' or 'add up to'.

a) In a series circuit, the current at all points … the current taken from the supply.
b) The voltages across each component in series … the voltage of the supply.
c) The currents in all of the parallel branches … the current taken from the supply.
d) The voltage across every component in parallel … the voltage across every other component.

SAQ 17 When a lot of appliances are connected to one socket using a multiway adaptor, they are connected in parallel.

Explain why this is dangerous.

C For Credit Level only, you should be able to draw circuit diagrams to show how car lights are wired. The following SAQ will give you practice in using the correct symbols as well as drawing circuit diagrams.

Question

C **SAQ 18** A car has four 12 V sidelights, two 12 V headlights and a 12 V battery. Two switches control the lights. The main switch controls all the lights and the headlight switch controls the headlights only when the main switch is on. Draw the circuit diagram, using the correct symbols.

C When resistors are connected together in a circuit, the combined resistance can be calculated. The relationship to use depends on whether the resistors are connected in series or in parallel.

Question

SAQ 19 What is the relationship to use to find the combined resistance, R_T, of two resistors R_1 and R_2 connected

a) in series
b) in parallel?

Examples

A 33 Ω resistor is connected in series with a 47 Ω resistor.

Calculate the combined resistance.

Solution

$R_1 = 33\,\Omega$
$R_2 = 47\,\Omega$
$R_T = ?$

$$R_T = R_1 + R_2$$
$$= 33 + 47$$
$$= 80\,\Omega$$

A 4 Ω resistor is connected in parallel with a 12 Ω resistor.

Calculate the combined resistance.

Solution

$R_1 = 4\,\Omega$
$R_2 = 12\,\Omega$
$R_T = ?$

$$\frac{1}{R_T} = \frac{1}{R_1} + \frac{1}{R_2}$$
$$= \frac{1}{4} + \frac{1}{12}$$
$$= \frac{4}{12}$$
$$R_T = \frac{12}{4}$$
$$= 3\,\Omega$$

Household circuits

All the appliances that you plug in to electrical sockets are connected in parallel by the household wiring. Mains fuses, or sometimes **circuit breakers**, protect the mains wiring in a house. A circuit breaker is an automatic switch that can be used instead of a fuse. A circuit breaker quickly stops the current if it becomes too big.

For Credit Level only, you need to know more about household wiring. Figure 2.7 gives some information. Use the diagram to answer the following SAQ.

Question

SAQ 20 a) Describe a ring circuit.
b) Give two advantages of using a ring circuit rather than a simple parallel circuit to supply power sockets.
c) Give two differences between the lighting circuit and the power ring circuit.
d) Give one reason why a circuit breaker is better than a fuse.

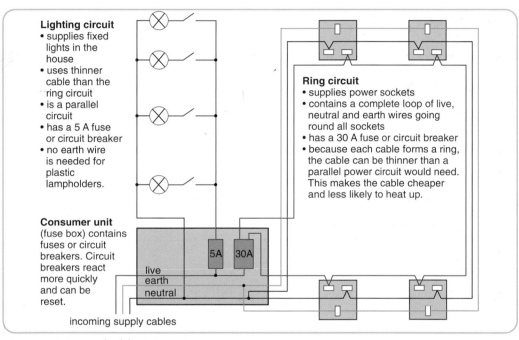

Lighting circuit
• supplies fixed lights in the house
• uses thinner cable than the ring circuit
• is a parallel circuit
• has a 5 A fuse or circuit breaker
• no earth wire is needed for plastic lampholders.

Ring circuit
• supplies power sockets
• contains a complete loop of live, neutral and earth wires going round all sockets
• has a 30 A fuse or circuit breaker
• because each cable forms a ring, the cable can be thinner than a parallel power circuit would need. This makes the cable cheaper and less likely to heat up.

Consumer unit (fuse box) contains fuses or circuit breakers. Circuit breakers react more quickly and can be reset.

5A   30A

live
earth
neutral

incoming supply cables

Figure 2.7 Household wiring

What You Should Know

◆ About current and voltage in series and parallel circuits.
◆ Why connecting too many appliances to one socket is dangerous.
◆ About practical applications needing two switches in series.
◆ How to draw circuit diagrams for car lighting circuits.
◆ How to do calculations involving the relationships for resistors connected in series and in parallel.
◆ How house wiring connects appliances.
◆ About mains fuses and circuit breakers and, for Credit Level only, why a circuit breaker is better than a fuse.
◆ The advantages of a ring circuit and how a ring circuit is different from a lighting circuit.

Answers

SAQ 16

a) is the same as

b) add up to

c) add up to

d) is the same as

SAQ 17

A large current could be taken from the mains supply.

SAQ 18

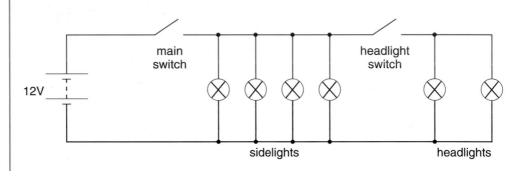

Figure 2.8

SAQ 19

a) Series: $R_T = R_1 + R_2$

b) Parallel: $\dfrac{1}{R_T} = \dfrac{1}{R_1} + \dfrac{1}{R_2}$

SAQ 20

a) A ring circuit contains a complete loop of live, neutral and earth wires going round all sockets.

b) Any two from: the cable for a ring circuit is thinner and cheaper; there is less current in each cable; a ring circuit is convenient because there can be a lot of power sockets.

c) Any two from: the lighting circuit supplies the fixed lights, the ring circuit supplies the power sockets; the lighting circuit uses thinner cable; the lighting circuit is a simple parallel circuit; the lighting circuit has a 5 A fuse, the ring circuit has a 30 A fuse; the lighting circuit does not need an earth wire if plastic lampholders are used.

d) Either a circuit breaker works faster than a fuse when the current becomes too high or a circuit breaker can be reset by pressing a switch.

USING ELECTRICITY

The magnetic effect

There is a magnetic field in the space around a wire that has a current in it. This is known as the **magnetic effect** of a current. If the wire is wound into a coil and an iron core is placed in the centre of the coil, an **electromagnet** or a **solenoid** is made.

When a wire that has a current in it is placed in a magnetic field, a force acts on the wire. This effect is used in the **electric motor**.

You may have tried to make a simple electric motor in class, although you may not have been too successful – it's very fiddly to get a simple motor to work. Your motor may have looked like the one in Figure 2.9.

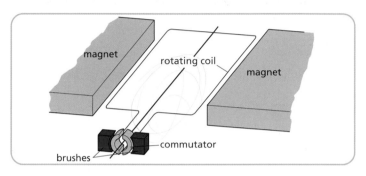

Figure 2.9 A simple electric motor

Questions

C **SAQ 21** What two things does the direction of the force on a wire carrying a current depend upon?

SAQ 22 Explain how an electric motor works. Your explanation should be based on the forces acting on the coil and should say what the purpose of the brushes and the commutator is.

SAQ 23 Commercial motors are different in some ways from the simple motor you may have made in class. Copy and complete the table to say why there are these differences.

	Simple motor	Commercial motor	Reason for difference
brushes	wires	carbon blocks	
commutator and coils	two section commutator and single coil	many section commutator in a ring and a coil for each pair of sections opposite each other round the ring	
magnet	permanent magnet	electromagnet (called field coils)	

HOW TO PASS STANDARD GRADE PHYSICS

What You Should Know

- ◆ About the magnetic effect of a current and some applications of it.
- ◆ About the force on a wire carrying a current in a magnetic field.
- ◆ The parts of an electric motor.
- C ◆ How an electric motor works.

Answers

SAQ 21

The direction of the current and the direction of the field.

SAQ 22

(Refer to the diagram of the simple electric motor.)

When the coil is in the horizontal position, the brushes and commutator supply current to the coil. The current in opposite sides of the coil is in opposite directions. This makes the coil rotate. When the coil moves a little beyond the vertical position, the brushes make contact with the opposite sides of the commutator. This means that whatever side of the coil is at the left always has a force in the same direction (and in the opposite direction to the side at the right). So the coil continues to rotate.

SAQ 23

	Simple motor	Commercial motor	Reason for difference
brushes	wires	carbon blocks	good electrical contact ◆ reduces wear on commutator ◆ allows free rotation
commutator and coils	two section commutator and single coil	many section commutator in a ring and a coil for each pair of sections opposite each other round the ring	◆ rotation is smoother ◆ continuous force because one coil is always in operation
magnet	permanent magnet	electromagnet (called field coils)	◆ magnetic field can be stronger for the same size of magnet ◆ motor can operate using alternating current

Chapter 3

HEALTH PHYSICS

Physics is used in medicine to diagnose and treat patients. You will be familiar with **thermometers** to measure temperature. Probably a doctor will have used a **stethoscope** to listen to sounds from inside your body. You may wear glasses or contact lenses. If not, you will certainly know someone who does. Other areas of Physics that are used in medicine are **ultrasonics**, **fibre optics** and **radioactivity**.

All of these areas are covered in this chapter on Health Physics.

Thermometers

A thermometer is an instrument that is used to measure temperature.

Question

SAQ 1 Write out the following statements about different types of thermometer, matching the first and the last parts of the statements.

a) A liquid in glass thermometer contains a liquid that...

b) A liquid crystal thermometer used crystals that...

c) An electronic thermometer contains a component that...

...has an electrical property (e.g. resistance) that changes as the temperature changes.

...expands in volume as the temperature increases.

...change colour as the temperature changes.

Doctors use clinical thermometers to measure body temperature. Normal body temperature is 37 °C. A body temperature only a few degrees higher or lower than this value is an indication that the person is ill.

There are three main differences between a clinical thermometer and an 'ordinary' thermometer.

1   A clinical thermometer has a means of maintaining the reading after it has been removed from the body.

2   A clinical thermometer measures only a narrow range of temperatures around normal body temperature.

3   A clinical thermometer is more sensitive to small changes in temperature – it can read accurately to 0·1 of a degree Celsius.

Question

SAQ 2   Describe how each of the three differences above is included in a liquid in glass clinical thermometer.

To measure body temperature using a clinical thermometer, the doctor must first make sure the thermometer is ready. The thermometer is then placed in contact with the body for long enough to indicate the temperature of the body. The thermometer is removed from the body and the scale read.

What You Should Know

◆   What a thermometer measures and how it does so.
◆   How a liquid in glass thermometer operates.
◆   The differences between a clinical and an 'ordinary' thermometer.
◆   How and why a clinical thermometer is used to measure body temperature.

Answers

SAQ 1

a)   A liquid in glass thermometer contains a liquid that expands in volume as the temperature increases.

b)   A liquid crystal thermometer uses crystals that change colour as the temperature changes.

c)   An electronic thermometer contains a component that has an electrical property (e.g. resistance) that changes as the temperature changes.

SAQ 2

1   A constriction is made in the glass bore, near to the bulb. This constriction stops the liquid flowing back down when removed from the body.

2 and 3   The clinical thermometer has a very narrow glass bore. The scale measures from 35 °C to 43 °C in 0·1 °C steps. Very small temperature changes cause large changes in the length of the liquid column.

Sound

The stethoscope

A stethoscope is a type of hearing aid that is used by doctors to listen to sounds that come from the body – usually the heart and the lungs.

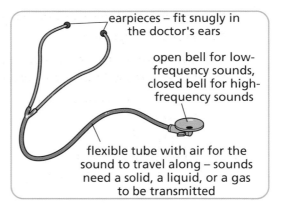

earpieces – fit snugly in the doctor's ears

open bell for low-frequency sounds, closed bell for high-frequency sounds

flexible tube with air for the sound to travel along – sounds need a solid, a liquid, or a gas to be transmitted

Figure 3.1 The stethoscope

HEALTH PHYSICS

Questions

SAQ 3 Explain how a stethoscope is used to listen to sounds from a body.

SAQ 4 The lungs give out high frequency sounds. The heart gives out lower frequency sounds. Use information from Figure 3.1 to say which bell is used for each of these sounds.

Ultrasound

Humans can hear sounds over a certain range of frequencies. High frequency vibrations that are beyond the range of human hearing are called **ultrasounds**.

Question

SAQ 5 What is taken as the range of human hearing?

Hints and Tips

There are several points that the examiner is looking for in a seemingly simple question like SAQ 5.

◆ When you are asked to give a range, make sure you give two values, not just a single value.

◆ Always remember to include the unit.

◆ You may have carried out an experiment in class on your range of hearing. The examiner does not want to know *your* range of hearing but the accepted range for all humans.

This means that you have to read a question carefully and answer what is asked, not what you would have liked to be asked.

You should be able to give an example of the use of ultrasound in medicine. The safest example to give is the use of ultrasound to create images of an unborn baby.

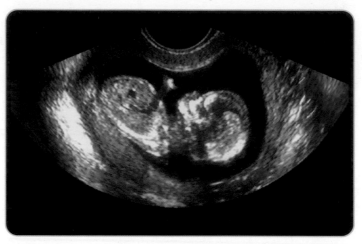

A scan produced using ultrasound

Hints and Tips

This is another situation where you have to be careful to write exactly what you mean. 'To take a picture of a baby' is too loose because you don't need ultrasound to see a baby after it is born ('unborn' is necessary). 'To take a picture of a baby in its mother's stomach' will make the examiner cringe and will not gain you any marks. Babies are never in their mother's stomachs – they develop in the womb.

Question

C **SAQ 6** Explain how ultrasound can be used to create images of the inside of a body. Key words that you could use include reflect, echo, body tissue.

Noise

We are constantly surrounded by sound. Unwanted sound is called **noise**. Sound levels are measured on the **decibel** (dB) scale. Too much noise over a period of time can damage hearing.

Question

SAQ 7 Complete the table of sound levels in the range 0 dB to 120 dB, by entering the following sources of noise: disco loudspeakers about 1 metre away, motorway traffic nearby, normal conversation in a room, a watch ticking.

Noise level (dB)	Source of noise
0	silence (threshold of hearing)
30	
60	
90	
120	
140	threshold of pain

What You Should Know

- What is needed for sounds to be transmitted.
- How a stethoscope works.
- What ultrasounds are.
- A use of ultrasound in medicine.
- About noise pollution.
- Examples of sound levels.

Answers

SAQ 3

The bell of the stethoscope is placed on the patient's body. This gathers the sounds from the body and amplifies them. The sounds are transmitted along the air in the flexible tube to the earpieces. The earpieces are a tight fit in the doctor's ears to exclude other sounds.

SAQ 4

Sounds from the lungs – closed bell.   Sounds from the heart – open bell.

SAQ 5

The range of normal human hearing is taken as 20 Hz to 20 000 Hz.

Answers

SAQ 8

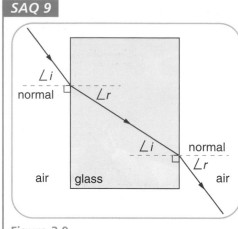

air   glass   air

Figure 3.8

SAQ 9

∠i
normal   ∠r

∠i   normal
∠r

air   glass   air

Figure 3.9

SAQ 10

a) A convex lens is thicker in the centre and thinner at the edge.
b) A concave lens is thinner in the centre and thicker at the edge.
c) A convex lens brings parallel rays of light to a focus (it converges them).
d) A concave lens spreads out parallel rays of light (it diverges them).
e) A more powerful convex lens is more curved than a less powerful one. (Sometimes this is described as being 'fatter'.)
f) A more powerful convex lens changes the direction of rays of light more than a less powerful lens. Because of this, parallel rays of light are brought to a focus closer to the lens. A more powerful lens has a shorter **focal length** – the distance from the centre of the lens to the focus.

SAQ 11

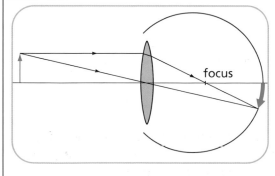

focus

Figure 3.10

SAQ 12

a) long sight and normal sight
b) long sight
c) short sight
d) long sight
e) short sight
f) short sight and normal sight
g) normal sight

ABERDEEN COLLEGE
GALLOWGATE LIBRARY
01224 612138

Answers *continued*

SAQ 13

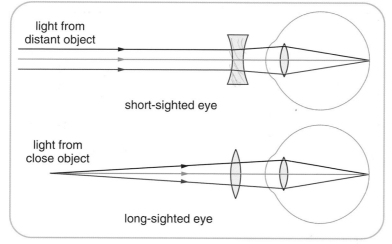

Figure 3.11

SAQ 14

a) The fibre optics bundle transmits the light but not the heat produced by the light source to the inside of the body.

b) Light is transmitted along the fibre optics in an endoscope by total internal reflection.

c) One fibre optics bundle transmits cold light to illuminate the inside of the body. The reflected light is then transmitted up the other fibre optics bundle so that an image can be viewed.

Using the spectrum

Lasers, X-rays, ultraviolet radiation and infrared radiation are all used in medicine.

Lasers

A laser is a very intense beam of light. Because a laser concentrates a lot of energy in a small area, it can be used to perform delicate surgery. The laser seals blood vessels as it cuts them, so a laser is often called a 'bloodless scalpel'.

Question

SAQ 15 Choose one of the following areas of medicine. Describe how the laser is used in this application.

◆ Eye surgery
◆ Removing birth marks
◆ Destroying tumours

X-rays

X-rays are very high frequency waves that darken photographic film. They are absorbed by bone and are transmitted through less dense tissue.

C For Credit Level only, you should know about **computerised tomography**. (This imposing name is often referred to as a CAT scan.) With this process, the patient is placed inside a circle of X-ray detectors. An X-ray generator takes a lot of

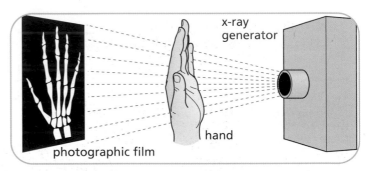

Figure 3.12 **Using X-rays in medicine**

images (about 480) of the patient as it moves around the circle in the same plane. The X-ray generator then repeats this process in different planes. In this way a three-dimensional, 360° image of the body can be created. A more detailed image can be built up and organs are not obscured.

Ultraviolet and infrared

Ultraviolet and infrared are two other types of radiation that are used in medicine.

Question

SAQ 16 The answers to all the parts of this question are either ultraviolet radiation or infrared radiation.

a) Used to create a thermogram – a special photograph that uses the radiation given off by a hot body.
b) Has a frequency higher than visible radiation.

Question continued ➢

Question *continued*

c) Too much exposure can cause skin cancer.
d) Used by physiotherapists to speed up the healing of strained muscles and tissue.
e) Has a frequency lower than visible radiation.
f) Used to treat skin conditions such as acne.
g) Used to sterilise equipment by killing harmful bacteria.
h) Used to diagnose tumours because they emit more heat than other tissue.
i) Helps to make vitamin D.

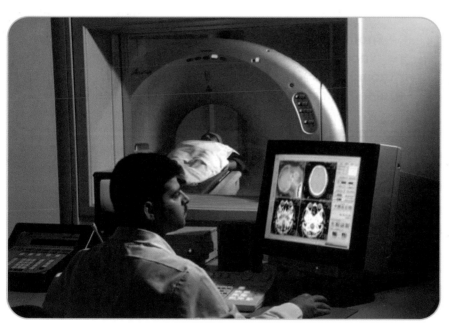

A CAT scan taking place

What You Should Know

- How a laser is used in medicine.
- How X-rays can be detected.
- How X-rays are used in medicine.
- About computerised tomography.
- How ultraviolet radiation is used in medicine.
- How infrared radiation is used in medicine.

HOW TO PASS STANDARD GRADE PHYSICS

Equivalent dose is measured in **sieverts** (Sv) or **millisieverts** (mSv). A dose of 1 mSv from any type of radiation would have exactly the same effect on the same tissue as a 1 mSv dose from another type of radiation.

What You Should Know

◆ How to draw a simple model of an atom.
◆ About the three types of radiation – alpha, beta and gamma radiation.
◆ About ionisation.

C ◆ Effects of radiation on materials and, for Credit Level only, how these effects are used to detect radiation.
◆ About the activity of a radioactive source.

C ◆ What is meant by the half-life of a radioactive source and how to measure it.
◆ How to do calculations to find the half-life of a radioactive element.
◆ The effects of radiation on living cells.
◆ Medical uses of radiation.
◆ Safety precautions needed when dealing with radioactive substances.
◆ About equivalent dose.

Answers

SAQ 17

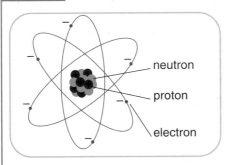

neutron

proton

electron

Figure 3.16

SAQ 18

a) a thin sheet of paper
b) a few millimetres of aluminium
c) several centimetres of lead
d) alpha radiation
e) gamma radiation

SAQ 19

Ionisation: A Geiger-Müller tube contains a low pressure gas. Radiation that enters the thin window ionises the gas and this causes a current between the two electrodes. This current pulse is recorded on a counter.

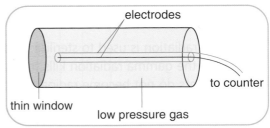

electrodes

to counter

thin window   low pressure gas

Figure 3.17 A Geiger-Müller tube

Answers *continued*

Fogging a photographic film:
A radiation film badge contains a small piece of photographic film behind various thicknesses of different absorbers. The badge is pinned onto clothing for a period of time. When the film is developed, the amount of fogging gives an indication of the person's exposure to radiation.

Scintillations: When radiation is absorbed by some materials, the energy causes small flashes of light – scintillations. The effect is used in a gamma camera.

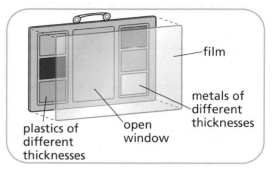

Figure 3.18 A radiation film badge

SAQ 20

Measure the background count rate using a Geiger-Müller tube connected to a counter. Measure the count rate at regular intervals with the source present for a period of time. Subtract the background count rate from each count rate with the source present to give the corrected count rate. Plot a graph of corrected count rate of the source against time. From the graph, measure the time taken for the corrected count rate of the source to drop by half. Repeat the measurement from the graph and calculate the average value of time taken – the half-life.

SAQ 21

1 Handling
2 Handling
3 Handling
4 Storing
5 Storing
6 Handling
7 Storing

ELECTRONICS

This unit of work is about electronics, but instead of studying the behaviour of individual components it covers **systems electronics**, where an electronic circuit is thought of as an **input part**, a **process part** and an **output part**. The input and the output parts are considered in detail, and then systems designed to carry out particular jobs are studied. We will mainly consider **digital** systems in this chapter, but some **analogue** systems are included.

Systems electronics

Systems electronics is a way of studying electronics without having to consider the components individually. All electronic systems can be considered as an input part, a process part and an output part, with a signal passing from one part to the next.

Figure 4.1 Block diagram of an electronic system

The input part takes information in and produces an electrical signal. This signal is processed or changed in some way by the process part. The output part converts the processed electrical signal into some form of energy.

Question

SAQ 1 Consider a portable radio as an electronic system. What are a) the input, b) the process and c) the output parts of this system?

There are many ways of sorting electronic systems, but one of the most useful is by considering the type of signal that the system deals with.

◆ An analogue signal is one that varies in a continuous way, and can take any value between a maximum and a minimum.

◆ A digital signal can have only particular set values. For Standard Grade Physics you only need to think of digital signals as being on or off, high or low. This is a **two-state** or **binary system**.

Question ?

SAQ 2 Label the two waveforms shown as either analogue or digital.

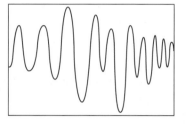 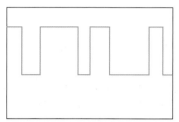

_____ waveform   _____ waveform

Figure 4.2

Electronic systems can have analogue or digital outputs.

Question ?

SAQ 3 Label the outputs from the electronic systems shown as either analogue or digital.

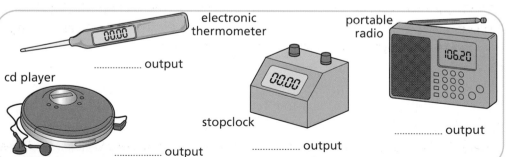

electronic thermometer
.... output

portable radio
.... output

cd player
.... output

stopclock
.... output

Figure 4.3 Electronic systems

What You Should Know ✓

◆ About the parts of an electronic system.
◆ Digital and analogue signals and outputs.

HOW TO PASS STANDARD GRADE PHYSICS

Answers

SAQ 1

a) The aerial is the input part.
b) The electronic circuits are the process part.
c) The loudspeaker is the output part.

SAQ 2

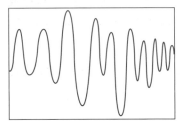
analogue waveform

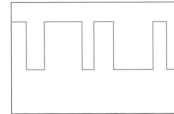
digital waveform

Figure 4.4

SAQ 3

Electronic thermometer – digital
Portable radio – analogue

CD player – analogue
Stopclock – digital

Input devices

Input devices fall into two groups:

◆ Devices that transfer energy to electrical energy and so provide an input signal – the **microphone**, the **thermocouple** and the **solar cell**.

◆ Devices that can be used to change a voltage, either on their own or as part of a **voltage divider** – the **thermistor**, the **light dependent resistor** (LDR), the **switch** and the **capacitor**.

Question

SAQ 4 Write out the following descriptions of the actions of the input devices mentioned above, along with the name of the device.

Question continued ➢

Question *continued*

    a) The voltage across this device increases with time during charging.
    b) This device has a resistance that decreases with increasing light intensity.
    c) Sounds are converted into electrical signals by this input device.
    d) This device converts light into electrical signals.
    e) A device that can be either on or off.
    f) This device has a resistance that changes with temperature.
    g) Heat is converted to electrical signals by this device.

The thermistor and the LDR are both types of resistors so the relationship $R = \dfrac{V}{I}$ applies to them.

Example

An LDR is in a circuit with 5 V across it. When placed in a dark room, the current through the LDR is 20 mA.

a) Show that the resistance of the LDR under these conditions is 250 Ω.
b) The LDR is taken to a brightly lit room. Suggest a value for the resistance of the LDR now.

Solution

a) $V = IR$

    $\therefore I = \dfrac{V}{R}$

       $= \dfrac{5}{20 \times 10^{-3}}$

       $= 250 \ \Omega$

b) Because the resistance of an LDR decreases as the light intensity increases, the resistance will be less than 250 Ω, say 200 Ω.

Hints *and* **Tips**

You will have noticed that part a) of the above example starts with the words 'Show that' and that the answer (250 Ω) is given. This is a form of wording that is sometimes used in the exam. The examiner does not use these words to confuse you – he uses them to help you.

Hints and *Tips* continued ➤

> **Hints and Tips continued**
>
> Suppose part a) read 'Calculate the resistance of the LDR.' If you couldn't answer
> this, you would also not be able to do part b), because you would not have a value
> of resistance to compare with. By using these words, and giving you the value of
> 250 Ω, the examiner has given you the chance to try the rest of the question, and
> so not lose all the marks, even although you might not be able to do part a).
> If you meet this type of question, treat it as an ordinary calculation question and be
> glad that you have been given the answer as a bonus!

C Voltage dividers

It is often useful to divide up a voltage either in a fixed ratio or as some quantity
changes (time, temperature or light level, for example). A voltage divider circuit
does this.

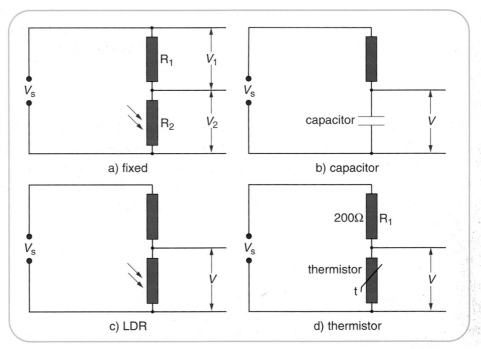

Figure 4.5 Voltage divider circuits

Consider the fixed voltage divider circuit in Figure 4.5a. Resistors R_1 and R_2 are in
series, so the same current is in them both.

$$I = \frac{V_1}{R_1} = \frac{V_2}{R_2}$$

$$\therefore \frac{V_1}{V_2} = \frac{R_1}{R_2}$$

ELECTRONICS

Questions

SAQ 5 a) After the circuit is switched on, what happens to the voltage *V* across the capacitor in Figure 4.5b, as time goes on?
b) How could the time to charge the capacitor in Figure 4.5b be changed? (This needs a change to the value of a component.)

SAQ 6 What happens to the voltage *V* across the LDR in Figure 4.5c when the LDR is taken from dark conditions into light conditions?

SAQ 7 At a temperature of 50°C the resistance of the thermistor in Figure 4.5d is 300 Ω, and the voltage across R_1 is 2 V.

Calculate the output voltage *V* at this temperature.

You should be able to suggest a suitable input device for a particular application. For General Level this is easier than for Credit Level because you are given a list to choose from. Try the following SAQ. If you are only doing General Level, you can look back to the beginning of this section to remind yourself of suitable input devices. For Credit Level, try to do the question without looking back.

Question

SAQ 8 What would be suitable input devices in the following situations?

a) As a time delay before switching off an automatic light.
b) The input device for a public announcement system at an airport.
c) An energy source for a spacecraft on a mission to other planets.
d) To monitor the temperature in a greenhouse.

What You Should Know

◆ Energy transformations in some input devices.
◆ About the thermistor, the LDR and the capacitor.
◆ How to carry out calculations with voltages and resistances in a voltage divider.
◆ About choosing suitable input devices.

Answers

SAQ 4

a) capacitor	d) solar cell	g) thermocouple
b) LDR	e) switch	
c) microphone	f) thermistor	

Answers *continued*

SAQ 5

a) The voltage increases.
b) The capacitance and/or the resistance could be changed. (The time to charge the capacitor increases as the capacitance and/or the resistance increases.)

SAQ 6

As the LDR is taken from dark to light, its resistance decreases. This means that its share of the fixed voltage V_s across the voltage divider goes down.

SAQ 7

$$\frac{V_1}{R_1} = \frac{V}{R_{thermistor}}$$

$$\therefore V = \frac{V_1}{R_1} \times R_{thermistor}$$

$$= \frac{2}{200} \times 300$$

$$= 3\,V$$

SAQ 8

a) capacitor
b) microphone
c) solar cell
d) thermistor or thermocouple

Output devices

The output device in an electronic system is used to convert the processed electrical signal into some useful type of energy, usually light, sound or movement.

Some of the output devices you may know include: buzzer, LED, lamp, 7-segment display, loudspeaker, electric motor, relay, solenoid.

Questions

SAQ 9 Choose a suitable output device for the following electronic systems:

    a) an automatic floodlight
    b) a calculator display
    c) central locking for car doors
    d) a radio
    e) a power-on indicator in a portable TV
    f) to turn a CD in a computer drive
    g) an audible electronic alarm.

Questions *continued* ➤

Questions *continued*

SAQ 10 Complete the key to output devices by entering the names of the devices listed on page 72 in the correct boxes.

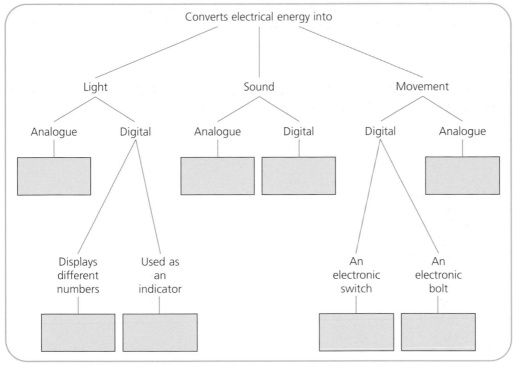

Figure 4.6 Key to output devices

The LED

Lamps give out light when they operate. They can be connected either way round in a circuit. There is another light-emitting output device, one that takes less current – the light emitting diode or LED. The symbol for an LED is shown in Figure 4.7.

An LED will only light if it is connected the correct way in a circuit. (The bar on the symbol goes to the negative end of the power supply.) Most LEDs only need about 2 V across them. Since most power supplies are more than 2 V, LEDs usually need a series resistor to protect them.

Figure 4.7 The circuit symbol for a light emitting diode (LED)

HOW TO PASS STANDARD GRADE PHYSICS

Example

C The LED in the circuit shown is designed to operate correctly with a voltage of 2·2 V across it and a current of 10 mA through it

Calculate the resistance of R, if the supply voltage is 4 V.

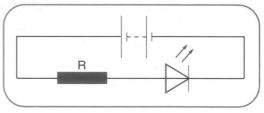

Figure 4.8

Solution

The LED and R are in series so the battery voltage splits up between them.

$$4\cdot0 = V_R + V_{LED}$$
$$\therefore V_R = 4\cdot0 - V_{LED}$$
$$= 4\cdot0 - 2\cdot2$$
$$= 1\cdot8\,V$$

The current is the same through the resistor and the LED.

$$R = \frac{V_R}{I}$$

$$= \frac{1.8}{10 \times 10^{-3}}$$

$$= 180\,\Omega$$

7-segment displays are used on many electronic devices, such as calculators and digital watches. They consist of seven bar-shaped LEDs (or more commonly these days LCDs – **liquid crystal displays**) arranged in the pattern shown in Figure 4.9.

By activating (lighting in the case of an LED display) suitable bars or segments, all the digits from 0 to 9 can be produced. A bank of 7-segment displays can be used to display numbers with several digits.

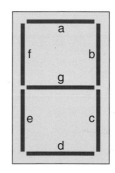

Figure 4.9 A 7-segment display

Question

SAQ 11 7-segment displays can also be used to display certain letters. Since the purpose of using this book is to pass Standard Grade Physics, which segments would be activated on a bank of four 7-segment displays to spell the word 'PASS'?

What You Should Know

◆ Energy transformations in some output devices.
◆ About digital and analogue output devices.
◆ Choosing suitable output devices.
◆ About the LED.
◆ About the 7-segment display.

Answers

SAQ 9

a) lamp
b) 7-segment display
c) solenoid
d) loudspeaker
e) LED
f) electric motor
g) buzzer

SAQ 11

P – a, b, e, f, g
A – a, b, c, e, f, g
S – a, c, d, f, g
S – a, c, d, f, g

SAQ 10

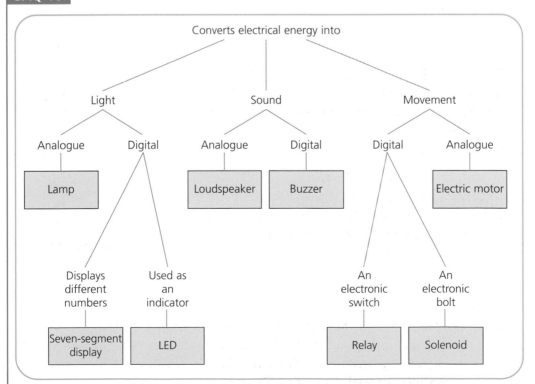

Figure 4.10

75

Digital processes

Transistors

The simplest process component is the **transistor**. A transistor can be used as a switch.

◆ When the transistor is non-conducting, it is like an open switch.
◆ When the transistor is conducting, it is like a closed switch – it is on.

The symbol for one type of transistor (the only one you need to know about) is shown in Figure 4.11.

Figure 4.11
The circuit symbol for a transistor

Hints and Tips

This is another popular question for the examiner to ask, 'Draw the symbol for a transistor.' Only about one student in 20 gets it correct. Make sure you are the one, not one of the other 19.

The transistor can be used in many switching circuits, for example:

◆ a fire alarm
◆ a burglar alarm
◆ an automatic nightlight or a parking light
◆ as a time-delay circuit.

Question

SAQ 12 Identify a use for the transistor switching circuit shown in Figure 4.12.

C SAQ 13 Explain how the circuit shown in Figure 4.12 operates.

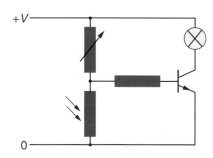

Figure 4.12 A transistor switching circuit

Digital logic gates

Digital logic gates are electronic circuits that have one or more inputs and usually one output. The inputs and outputs of logic gates are either high (when there is a high voltage at them, called 'logic 1') or low (a low voltage, or 'logic 0'). Whether the output of a gate is high (1) or low (0) depends on the type of gate and whether the input(s) is/are high or low.

Although there are lots of different types of logic gates, you only need to know about three:

◆ the two input AND gate ◆ the two input OR gate ◆ the NOT gate (inverter).

Logic gates can have one or more inputs. A truth table shows the output (1 or 0) for all possible combinations of inputs (1 or 0) to the gate. The names of the gates should help you to understand how they operate.

Remember

◆ The output of an AND gate is only high when one input AND the other input are both high.
◆ The output of an OR gate is high when one input OR the other input (OR both inputs) is high.
◆ The output of a NOT gate is NOT the same as its input.

Question

SAQ 14 a) Identify the following symbols for digital logic gates.

(i) _____ gate   (ii) _____ gate   (iii) _____ gate

Figure 4.13 Digital logic gates

b) Identify the gates that have the following truth tables.

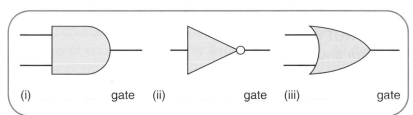

Input		Output
A	B	
0	0	0
0	1	1
1	0	1
1	1	1

Input		Output
A	B	
0	0	0
0	1	0
1	0	0
1	1	1

Input	Output
0	1
1	0

i) _____ gate   ii) _____ gate   iii) _____ gate

Digital logic gates can be combined to solve problems.

Finally, each 4 binary digit number (known for short as a 4-bit number) is converted by another electronic circuit into code for the equivalent decimal number. The circuit that does this conversion is known as a **binary decoder**. The output from the binary decoder is used to drive a 7-segment display to show the decimal number. The complete clock, counter, decoder and display is shown below as a block diagram.

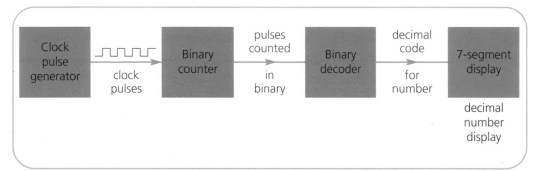

Figure 4.15 Block diagram of a clock, counter, decoder and display

What You Should Know

C
- ◆ About the transistor as an electronic switch.
- ◆ How to identify the purpose of a transistor switching circuit and, for Credit Level only, explain how a transistor switching circuit works.
- ◆ About the logic gates AND, OR and NOT.
- ◆ Logic voltage levels.
C
- ◆ The truth tables for AND, OR and NOT gates.
- ◆ How to use combinations of logic gates in simple situations.
C
- ◆ That a digital circuit can produce clock pulses and, for Credit Level only, how such a circuit works and how to change the frequency.
- ◆ About counter circuits.
C
- ◆ How to convert binary numbers to decimal numbers.

Answers

SAQ 12

The input part of this circuit is a voltage divider with an LDR. This shows that the circuit responds to changing light levels. The transistor acts to switch on the lamp (the output part). So the circuit could be used as an automatic nightlight that switches a lamp on when it becomes dark.

Answers continued

SAQ 13

The variable resistor and the LDR act as a voltage divider. As the light level falls, the resistance of the LDR increases. The voltage at the junction rises and switches on the transistor. When the transistor conducts, the lamp lights.

SAQ 14

a) i)   AND gate
   ii)   NOT gate
   iii) OR gate

b) i)   OR gate
   ii)   AND gate
   iii) NOT gate

SAQ 15

a)

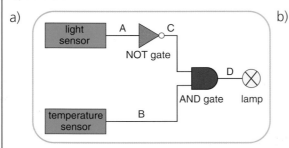

Figure 4.16

b)

A	B	C	D
0	0	1	0
0	1	1	1
1	0	0	0
1	1	0	0

SAQ 16

Initially the capacitor is uncharged and so the input to the NOT gate is low. The output from the NOT gate is high so the capacitor charges through the resistor. The voltage across the capacitor rises to high. The input to the NOT gate goes high, so its output goes low. The capacitor discharges through the resistor. This causes the input of the NOT gate to go low and its output to go high. This cycle repeats. The frequency of the clock pulses can be increased by reducing the capacitance of the capacitor and/or the resistance of the resistor.

SAQ 17

Decimal	Binary
0	0000
1	0001
2	0010
3	0011
4	0100
5	0101
6	0110
7	0111
8	1000
9	1001

HOW TO PASS STANDARD GRADE PHYSICS

Analogue processes

An **amplifier** increases the amplitude of an electronic signal. It boosts the strength of the audio signal from the input part. Amplifiers are used in radios, intercoms and music centres. Amplifiers can also boost other signals, such as video signals in a television or a video camera.

Question

SAQ 18 Which of the following devices contain an amplifier?

radio, toaster, intercom, torch, music centre, washing machine, television, microwave oven, video camera, refrigerator, MP3 player

The voltage gain of an amplifier is given by the relationship

$$\text{voltage gain} = \frac{\text{output voltage}}{\text{input voltage}} \qquad \text{or} \qquad V_{gain} = \frac{V_o}{V_i}$$

Example

The input voltage to an amplifier is 150 mV. The output voltage from the amplifier is 6 V. Calculate the voltage gain of the amplifier.

Solution
input voltage = 150 mV = 0·15 V
output voltage = 6 V
voltage gain = ?

$$\text{voltage gain} = \frac{\text{output voltage}}{\text{input voltage}}$$
$$= \frac{6}{0·15}$$
$$= 40$$

Hints and Tips

This is one instance where you will lose half a mark if you give a unit in your answer. Voltage gain is a ratio and so it does not have any units.

Question

SAQ 19 Figure 4.17 shows apparatus that could be used to measure the voltage gain of an amplifier.

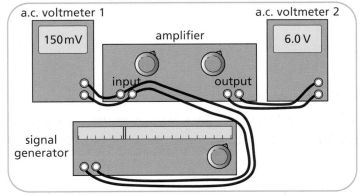

Figure 4.17 Measuring the voltage gain of an amplifier

Describe how to do so.

The power output of an amplifier is calculated using the relationship $P = \dfrac{V^2}{R}$ where P is the output power, V is the output voltage and R is the resistance or impedance of the circuit.

Example

An amplifier has an output voltage of 20 V. Calculate the power it delivers to a loudspeaker that has a resistance of 8 Ω.

Solution

output voltage = 20 V
resistance = 8 Ω
power = ?

$$P = \frac{V^2}{R}$$
$$= \frac{20^2}{8}$$
$$= 50\,\text{W}$$

The power gain of an amplifier is given by the relationship

$$\text{power gain} = \frac{\text{output power}}{\text{input power}} \qquad \text{or} \qquad P_{gain} = \frac{P_o}{P_i}$$

Example

C The input power to an amplifier is 200 mW. The output power from the amplifier is 50 W.

Calculate the power gain of the amplifier.

Solution

input power = 200 mW = 0·2 W
output power = 50 W

$$\text{power gain} = \frac{\text{output power}}{\text{input power}}$$

$$= \frac{50}{0·2}$$

$$= 250$$

What You Should Know

◆ About amplifiers – what they do and devices that use them.

◆ How to do calculations using the relationship $\text{voltage gain} = \dfrac{\text{output voltage}}{\text{input voltage}}$

C ◆ How to measure the voltage gain of an amplifier.

◆ How to do calculations using the relationship $P = \dfrac{V^2}{R}$.

◆ How to do calculations using the relationship $\text{power gain} = \dfrac{\text{output power}}{\text{input power}}$.

Answers

SAQ 18

The radio, intercom, music centre, television, video camera and MP3 player all contain amplifiers.

SAQ 19

Use the signal generator to supply an alternating voltage to the input of the amplifier. Measure this input voltage using a.c. voltmeter 1. Measure the output voltage using a.c. voltmeter 2. Calculate the voltage gain using

$$\text{voltage gain} = \frac{\text{output voltage}}{\text{input voltage}}.$$

Chapter 5

TRANSPORT

This chapter is about the movement or motion of objects. Movement is described in terms of **speed**, either **average speed** or **instantaneous speed**, as well as **acceleration** or how quickly speed changes. A lot of information about motion can be obtained from **speed–time graphs**.

It is **forces** that cause objects to move and forces that can stop movement. Included in this unit is a look at **Newton's laws** explaining how forces and motion are related. Moving objects have a particular type of energy – **kinetic energy**. This and other sorts of energy are covered along with a further look at power and work done.

Speed and acceleration

Average and instantaneous speed

The speed of an object is the distance the object moves divided by the time taken.

$$\text{speed} = \frac{\text{distance}}{\text{time}} \qquad v = \frac{s}{t}$$

This will give the average speed unless the time (and therefore distance) is very short.

Question

SAQ 1 Describe how to measure the average speed of a friend running in the playground. Use the relationship above. Decide what measurements you need to make and how you are going to make them.

Example

Your friend takes 12·5 s to run 50·0 m in the playground. Calculate the average speed.

Solution

time $t = 12\cdot5\,\text{s}$

distance $s = 50\cdot0\,\text{m}$

average speed $v = ?$

$$v = \frac{s}{t}$$

$$= \frac{50\cdot0}{12\cdot5}$$

$$= 4\cdot0\,\text{m/s}$$

A car speedometer records the speed of the car at every instant of time – instantaneous speed. Instantaneous speed is calculated using the same relationship, the difference being that the time is very short – the shorter the better.

To measure an instantaneous speed you need an electronic timer which is started and stopped by the object being timed. It is important that the scale divisions are fine enough. An electronic timer that is only precise to the nearest second would not usually be suitable for measuring instantaneous speeds. A hand-operated stopwatch would also be no use because reaction time to start and stop the stopwatch would be comparable to the time you are trying to measure.

Measuring instantaneous speed

The instantaneous speed of a vehicle on a runway can be measured using the apparatus shown in Figure 5.1.

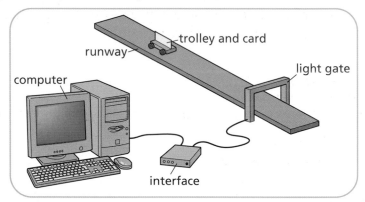

The length of a card attached to a trolley is measured. The timer is a computer and interface. The timer is started when the front of the card interrupts a light

Figure 5.1 Measuring instantaneous speed

beam at the light gate. The timer stops when the back of the card leaves the light beam. The instantaneous speed of the trolley through the light gate is given by the relationship $\text{instantaneous speed} = \dfrac{\text{length of card}}{\text{time recorded on timer}}$.

Acceleration

Unless an object is moving at a uniform speed, it will be speeding up or slowing down. We call the rate at which an object changes speed its acceleration.

$$\text{acceleration} = \frac{\text{change in speed}}{\text{time taken}} \qquad a = \frac{\Delta v}{t}$$

From this definition, you can see that the unit of acceleration is the unit of speed divided by the unit of time.

Question

SAQ 2 Write out the SI (Système International – the system of units that you should use for Standard Grade Physics) unit for acceleration, in full and in symbols.

Hopefully you wrote m/s² as your answer to SAQ 2. There are other ways of writing derived units like this one. Using what is called index notation, this unit can also be written as m s⁻². However, it is far better to stick to one way only and by doing so, you will not get confused. Use the solidus (/) or 'per' notation as that is the way the unit will be written in the Credit paper. If you try to remember other ways, you will be in danger of writing a combination of both which is wrong.

Acceleration is the rate at which speed changes. If the speed slows down, we would sometimes refer to this as **deceleration**, but it is still an acceleration. The value would have a negative sign.

Example

A car increases its speed by 9 m/s in 5 s. Calculate the acceleration of the car.

Solution

change in speed $\Delta v = 9$ m/s
time taken $t = 5$ s
acceleration $a = ?$

$$a = \frac{\Delta v}{t}$$

$$= \frac{9}{5}$$

$$= 1.8 \, \text{m/s}^2$$

Speed–time graphs

A speed–time graph is a way of showing the motion of an object over a period of time.

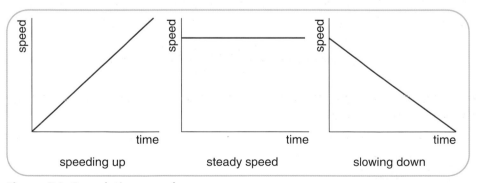

Figure 5.2 **Speed–time graphs**

Question

SAQ 3 Match the motions mentioned to the speed–time graphs shown in Figure 5.3 by drawing out the graphs and writing the motions underneath each.

a) A trolley released on a slope.
b) An aeroplane travelling at a steady speed across the Atlantic.
c) A car accelerating to overtake a lorry.
d) An athlete in a sprint race.
e) A driver seeing traffic lights change to red and braking to a stop.
f) A lift travelling between floors.

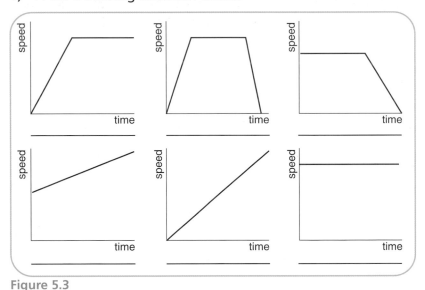

Figure 5.3

Acceleration can be calculated from a speed–time graph. At General Level you will only be asked to do this for motion with a single constant acceleration, but at Credit Level there could be a graph showing more than one constant acceleration. In either case, you find the change of speed during the correct time period and calculate the acceleration using the relationship $\text{acceleration} = \dfrac{\text{change in speed}}{\text{time taken}}$.

Example

The speed–time graph of a car accelerating away at traffic lights is shown in Figure 5.4.
Calculate the acceleration of the car.

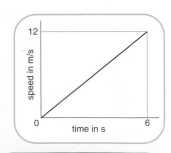

Figure 5.4

Example *continued* ➤

Example *continued*

Solution

change in speed $\Delta v = 12 - 0 = 12$ m/s

time for change $t = 6$ s

acceleration $a = ?$

$$a = \frac{\Delta v}{t}$$
$$= \frac{12}{6}$$
$$= 2\,\text{m/s}^2$$

C In addition, for Credit Level only you should be able to calculate distance gone from a speed–time graph. To do this you have to remember that the distance gone is represented on a speed–time graph by the area between the graph and the time axis (the area under the graph) for the correct time period.

Example

C The speed–time graph for a car travelling between two sets of traffic lights is shown in Figure 5.5.

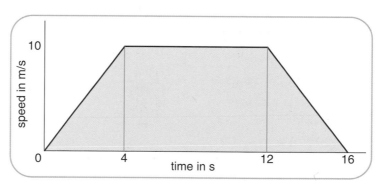

Figure 5.5

C a) Calculate the acceleration of the car during the period 12 s to 16 s.

b) Calculate the distance between the two sets of traffic lights.

Solution

a) change in speed $\Delta v = 0 - 10 = -10$ m/s

time for change $t = 4$ s

acceleration $a = ?$

$$a = \frac{\Delta v}{t}$$
$$= \frac{-10}{4}$$
$$= -2{\cdot}5\,\text{m/s}^2$$

b) distance between two sets of traffic lights = area under graph
   = area under 0–4 s + area under 4–12 s + area under 12–16 s
   = $(\frac{1}{2} \times 4 \times 10) + (8 \times 10) + (\frac{1}{2} \times 4 \times 10)$
   = $20 + 80 + 20$
   = 120 m

Hints and Tips

You will remember that a deceleration is a negative acceleration. Don't forget to include the negative sign in an answer like the one shown in part a) on page 89. If you say that the change in speed is +10 m/s, you will not be given any marks from the substitution onwards, because this is the wrong change.

C There is one further relationship that you should be able to use for Credit Level only. It involves the relationship between initial speed (u), final speed (v), time (t) and acceleration (a)

$$\text{acceleration} = \frac{\text{final speed} - \text{initial speed}}{\text{time}} \qquad a = \frac{v - u}{t}$$

Example

C To overtake a lorry, a car increases its speed from 13 m/s to 22 m/s in a time of 3 s.

Calculate the acceleration of the car.

Solution

initial speed $u = 13$ m/s
final speed $v = 22$ m/s
time $t = 3$ s
acceleration $a = ?$

$$a = \frac{v - u}{t}$$

$$= \frac{22 - 13}{3}$$

$$= \frac{9}{3}$$

$$= 3 \, \text{m/s}^2$$

What You Should Know

- What is meant by average speed and instantaneous speed.
- How to measure average and instantaneous speeds.
- How to do calculations using the relationship $\text{average speed} = \dfrac{\text{distance}}{\text{time}}$.
- What is meant by acceleration.
- How to do calculations using the relationship $\text{acceleration} = \dfrac{\text{change in speed}}{\text{time taken}}$.
- About speed–time graphs for different types of motion.
- **C** How to calculate acceleration and, for Credit Level only, distance gone from speed–time graphs.
- How to do calculations using the relationship
  $$\text{acceleration} = \frac{\text{final speed} - \text{initial speed}}{\text{time}}.$$

Answers

SAQ 1

Measure a suitable distance in the playground using a trundle wheel or a surveyor's tape. Use a stopwatch to time how long your friend takes to run this distance.

Calculate speed using $\text{average speed} = \dfrac{\text{measured distance}}{\text{time on stopwatch}}$.

SAQ 2

The SI unit of acceleration comes from the definition of acceleration – change of speed/time taken – so it is metre per second per second or m/s/s. This is usually written as m/s^2.

SAQ 3

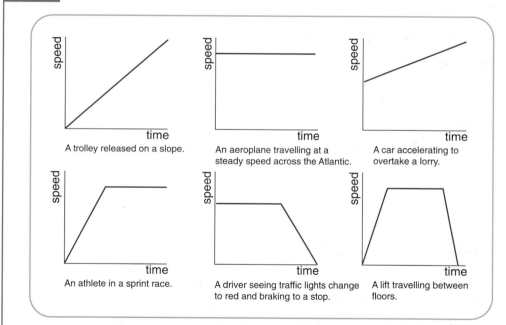

A trolley released on a slope.

An aeroplane travelling at a steady speed across the Atlantic.

A car accelerating to overtake a lorry.

An athlete in a sprint race.

A driver seeing traffic lights change to red and braking to a stop.

A lift travelling between floors.

Figure 5.6

Forces

Forces and movement

Forces cause changes in an object. They can change the shape of an object or they can change the movement of an object – either the speed or the direction of travel.

Question

SAQ 4 Write down an example of a force causing each of the changes mentioned above.

Forces can be measured using a **Newton balance**.

A Newton balance contains a spring that extends when a force is applied to it. The extension is proportional to the force. So the greater the force on the end of the spring, the greater the extension. A pointer on the spring moves over a scale marked in **newtons** (N), the unit of force.

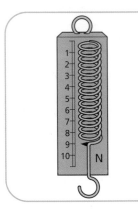

Figure 5.7 A Newton balance

Question

SAQ 5 A force of 2 N extends a spring from 100 mm to 140 mm. What length would the spring extend to with a force of 6 N applied?

Friction is a force that can oppose the movement of objects. When work is done against friction, energy is transferred as heat. Sometimes friction is useful to us and at other times it can be a nuisance and we have to reduce it as much as possible.

Question

SAQ 6 The following statements all relate to situations involving friction. For each statement, choose whether the force of friction has to be low or high. Also choose the correct ending for the statement. Write out all the statements in full.

a) When skiing, we want the force of friction to be as ... as possible, so we...
b) When holding a racket or a bat, the force of friction has to be as ... as possible, so we...
c) We want the force of friction in a car engine to be as ... as possible, so we...

Question continued ➤

Question *continued*

    d) We want the force of friction between the brake block and bicycle wheel to be as … as possible, so we…

    e) In a sports car, we want the force of friction due to air resistance to be as … as possible, so we…

    …can convert the kinetic energy of the bicycle into heat.
    …lubricate the moving parts with oil.
    …make the grips of rackets and bats from rubber.
    …streamline the shape of the car.
    …wax the bottom of the skis.

Mass and weight

Another force that you should know about is **weight**. Weight is the pull of the Earth on the **mass** of an object. Mass is measured in **kilograms** (kg).

Hints *and* **Tips**

Don't confuse mass and weight. Mass is how much 'stuff' or matter there is in an object. Mass is measured in kilograms. Weight is a force caused by the pull of the Earth on the matter that makes up the object. Weight is measured in newtons (N).

You can use a Newton balance to find the relationship between mass and weight. If you hang a mass of 1 kg on a Newton balance, you will see that the reading is approximately 10 N. So the pull of the Earth is approximately 10 N/kg.

Example

Calculate the weight of a bag of flour that has a mass of 1·5 kg.

Solution

mass = 1·5 kg   weight = mass × pull of Earth

pull of Earth = 10 N/kg   = 1·5 × 10

weight = ?   = 15 N

C For Credit Level only, you should know that the pull of the Earth on a mass is called the **gravitational field strength**.

$$\frac{\text{weight (N)}}{\text{mass (kg)}} = \text{gravitational field strength (N/kg)} \qquad \frac{W}{m} = g$$

This gives us the relationship $W = mg$.

Balanced and unbalanced forces

Equal-sized forces acting in opposite directions on an object are called **balanced forces**. Balanced forces are equivalent to no forces acting.

Question ?

SAQ 7 For each of the situations shown in Figure 5.8 say whether the forces acting are balanced or unbalanced.

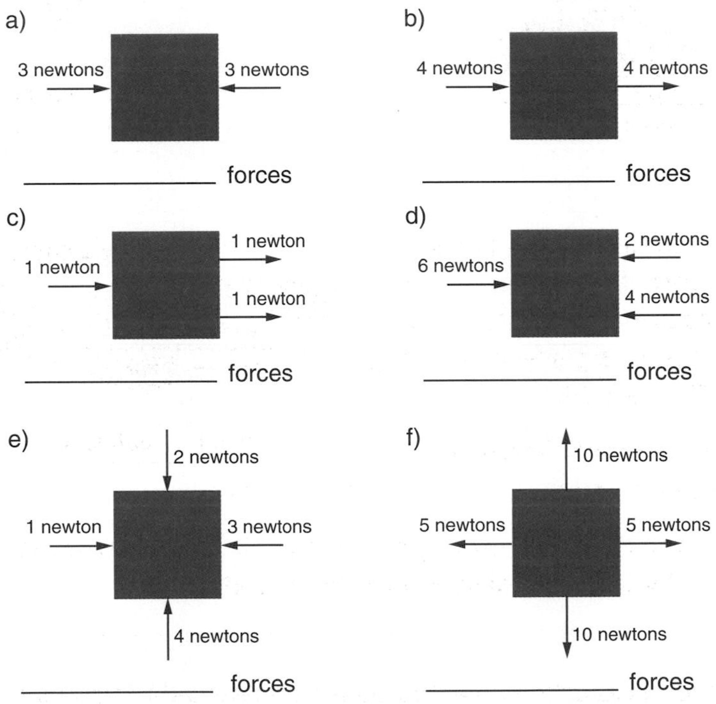

a)
3 newtons   3 newtons
_____ forces

b)
4 newtons   4 newtons
_____ forces

c)
1 newton
1 newton
1 newton
_____ forces

d)
2 newtons
6 newtons
4 newtons
_____ forces

e)
2 newtons
1 newton   3 newtons
4 newtons
_____ forces

f)
10 newtons
5 newtons   5 newtons
10 newtons
_____ forces

Figure 5.8

When balanced forces, or no forces at all, act on an object, its speed stays the same. This means that if the object is at rest, it will stay at rest; if it is moving, it will keep on moving at the same speed. This is known as **Newton's first law of motion**.

Questions

SAQ 8 For each of the situations shown in Figure 5.9 say whether the speed of the object stays the same or changes.

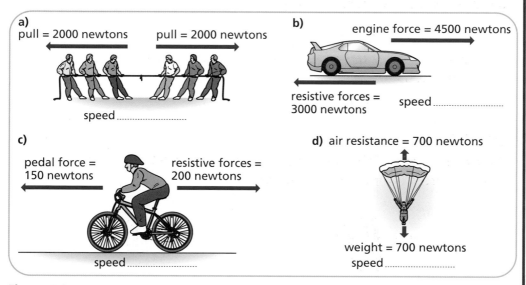

a)
pull = 2000 newtons   pull = 2000 newtons

speed ....

b)
engine force = 4500 newtons

resistive forces = 3000 newtons   speed ....

c)
pedal force = 150 newtons   resistive forces = 200 newtons

speed ....

d) air resistance = 700 newtons

weight = 700 newtons
speed ....

Figure 5.9

SAQ 9 An aircraft is flying at a uniform speed at a constant height. The forces acting are as shown in Figure 5.10.

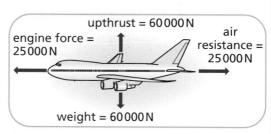

upthrust = 60 000 N
engine force = 25 000 N
air resistance = 25 000 N
weight = 60 000 N

Figure 5.10

Use Newton's first law of motion to explain

a) why the horizontal speed of the aircraft is constant
b) why the vertical speed of the aircraft is zero.

But what if the forces acting on an object are unbalanced? This situation is explained by **Newton's second law of motion**. This law relates the unbalanced force on an object and the mass of the object to the acceleration produced by the force.

◆ If the unbalanced force increases, the acceleration increases – acceleration varies directly as unbalanced force.

◆ If the mass increases, the acceleration decreases – acceleration varies inversely as mass.

These relationships are combined into the mathematical statement of Newton's second law of motion.

$$\text{unbalanced force (N)} = \text{mass (kg)} \times \text{acceleration (m/s}^2) \qquad F = ma$$

You should be able to do calculations using the relationship $F = ma$ at both General and Credit Levels. At Credit Level you may have to work out what the value of the unbalanced force is.

Example

The mass of the car in Figure 5.9b is 1000 kg. Calculate its acceleration.

Solution

mass $m = 1000\,\text{kg}$

unbalanced force $F = 4500 - 3000 = 1500\,\text{N}$

acceleration $a = ?$

$$a = \frac{F}{m}$$
$$= \frac{1500}{1000}$$
$$= 1 \cdot 5\,\text{m/s}^2$$

Hints and Tips

You will notice that in the above example, as in all the others in this book, I have not used what is sometimes called the 'triangle relationship'. I will break my rule just this once to show you what I mean.

The reason I don't use it is because I do not believe that using it helps you understand Physics. Also, if you write down a triangle relationship in the exam and then don't use it correctly, you will not be given the half mark for knowing the relationship. Better to learn the work by doing examples for yourself than trying to remember which symbol goes where in a triangle.

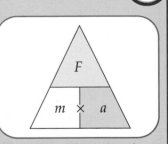

Figure 5.11 **The 'triangle relationship'**

Question

SAQ 10 Explain, by using Newton's laws of motion and in terms of the forces needed, why seatbelts are used in cars.

> ## What You Should Know
>
> - The changes that forces can cause in an object.
> - About the Newton balance and how it is used to measure force.
> - About the force of friction and the motion of objects.
> - What is meant by the mass and the weight of an object.
> - How to do calculations using the relationship **weight** = 10 × **mass**.
> C
> - What is meant by gravitational field strength.
> - About balanced and unbalanced forces.
> - About Newton's laws of motion.
> - How to do calculations using the relationship
>   **unbalanced force** = **mass** × **acceleration**.
> - Why seat belts are used in cars.

> ## Answers

SAQ 4

There are many examples you could have written. For example

- Pulling an elastic band causes its shape to change.
- Pushing a shopping trolley causes its speed to change.
- Hitting a tennis ball with a racket causes the direction of travel to change.

SAQ 5

A 2 N force extends the spring by 14 − 10 = 4 cm.

So a 6 N force extends the spring by $\frac{6}{2}$ × 4 cm = 12 cm to an extended length of 22 cm.

SAQ 6

a) When skiing, we want the force of friction to be as low as possible, so we wax the bottom of the skis.

b) When holding a racket or a bat, the force of friction has to be as high as possible, so we make the grips of rackets and bats from rubber.

c) We want the force of friction in a car engine to be as low as possible, so we lubricate the moving parts with oil.

d) We want the force of friction between the brake block and bicycle wheel to be as high as possible, so we can convert the kinetic energy of the bicycle into heat.

e) In a sports car, we want the force of friction due to air resistance to be as low as possible, so we streamline the shape of the car.

HOW TO PASS STANDARD GRADE PHYSICS

Answers continued

SAQ 7

a) balanced forces
b) unbalanced forces
c) unbalanced forces

d) balanced forces
e) unbalanced forces
f) balanced forces

SAQ 8

a) In the tug-of-war, the forces are balanced so the speed stays the same (stays at rest in this case).
b) The engine force of the car is greater than the resistive forces. The forces are unbalanced, so the speed changes.
c) The pedal force of the bicycle is less than the resistive forces. The forces are unbalanced, so the speed changes.
d) With the parachutist, the forces are balanced, so the speed stays the same.

SAQ 9

a) The engine force and the air resistance are balanced forces, so the horizontal speed remains constant.
b) The upthrust and weight are balanced forces so the vertical speed remains zero.

SAQ 10

When a car and its occupants are travelling at a steady speed, the forces are balanced (Newton's first law of motion). When the car brakes suddenly, or is involved in a collision, the forces acting on the car are unbalanced. This causes the car to decelerate, according to Newton's second law of motion.

If the occupants are not wearing seat belts, the forces on them remain balanced and they will continue to move forward at the same speed, probably hitting the windscreen (which is part of the car and so decelerating). If the occupant is wearing a seat belt, the belt will supply a force to slow the occupant down at the same rate as the car.

Movement means energy

Work

A session in the gym is often called a workout. This is because all the exercises in the gym involve doing work. When work is done, energy is transferred. When energy is transferred, a force is moved through a distance.

Work done is a measure of the energy transferred, and is calculated using the relationship

$$\text{work done (J)} = \text{force (N)} \times \text{distance (m)} \qquad E_w = Fs$$

Hints and Tips

There are different symbols used for work done, as well as kinetic energy and potential energy (both of which will be covered later in this section). Any of these symbols could be used in the exam. The symbols used are:

◆ Work done – W or WD or E_w (I will use E_w)
◆ Kinetic energy – KE or E_k (I will use E_k)
◆ Potential energy – PE or E_p (I will use E_p)

Example

A force of 125 N is used to move a box a distance of 3 m along the ground.

Calculate the work done by the force.

Solution

force $F = 125\,\text{N}$ $E_w = Fs$
distance $s = 3\,\text{m}$ $= 125 \times 3$
work done $E_w = ?$ $= 375\,\text{J}$

We are often interested in comparing the rate at which work is done or energy is transferred. Power is the rate of doing work or the rate at which energy is transferred.

$$\text{power (W)} = \frac{\text{work done (J)}}{\text{time (s)}} \quad \text{or} \quad \frac{\text{energy transferred}}{\text{time}} \qquad P = \frac{E}{t}$$

You met power in Chapter 2 in connection with electrical energy.

Example

The force in the previous example acts for a time of 5 s. Calculate the power involved.

Solution

work done (energy transferred) $E = 375\,\text{J}$ $P = \dfrac{E}{t}$
time $t = 5\,\text{s}$
power $P = ?$ $= \dfrac{375}{5}$
                                             $= 75\,\text{W}$

Work has to be done against gravity to lift an object up. In doing so, energy is transferred to the object in the form of **gravitational potential energy**. The increase in gravitational potential energy E_p in an object of mass m, raised through a height of h is given by the relationship

gravitational potential energy = mass × gravitational field strength × height

or $E_p = mgh$

When an object is lowered, its gravitational potential energy becomes less and work is done by gravity.

Example

A crate of mass 50 kg is raised a distance of 2 m from the ground onto the back of a lorry.

a) Calculate how much work is done.
b) How much extra gravitational potential energy does the crate have when on the lorry?
c) How much work is done by gravity when the crate is later lowered from the lorry to the ground?

Solution
a) mass $m = 50$ kg $E_w = Fs$
    distance $h = 2$ m $= mgh$
    work done $E_w = ?$ $= 50 \times 10 \times 2$
                                                $= 1000$ J

b) The gravitational potential energy transferred to the crate is the work done on the crate, so $E_p = 1000$ J.

c) The work done by gravity is the decrease in gravitational potential energy, so $E_w = 1000$ J.

Energy and movement

A moving object has kinetic energy. When you pedal your bicycle along a straight road at constant speed, you transform chemical energy in your muscles to heat energy overcoming the friction on the road and in the air.

Question

SAQ 11 Consider the four situations shown at different times during a bike journey.

Copy and complete the main energy transformations for each of the situations shown.

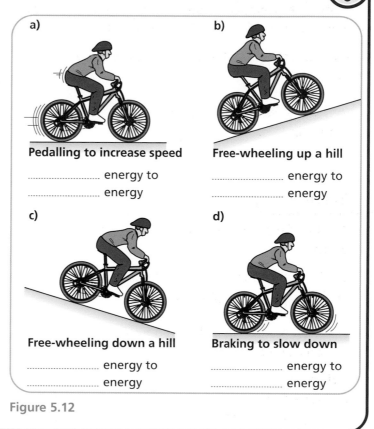

a)

Pedalling to increase speed

.... energy to

.... energy

b)

Free-wheeling up a hill

.... energy to

.... energy

c)

Free-wheeling down a hill

.... energy to

.... energy

d)

Braking to slow down

.... energy to

.... energy

Figure 5.12

The kinetic energy that an object has depends on its mass and its speed. The greater its mass, the greater its kinetic energy; the greater its speed, the greater its kinetic energy.

C For Credit Level only, you should be able to use the relationship between kinetic energy, mass and speed, $E_k = \frac{1}{2} mv^2$.

Hints and Tips

C Don't be confused by the square in this expression. It is only the v that is squared. Written out, the expression is $E_k = \frac{1}{2} \times m \times v \times v$. Also, don't forget to square the value of v when you do the calculation, otherwise you will lose most of the marks.

Example

C A car of mass 1200 kg is travelling at 13 m/s. Calculate its kinetic energy.

Solution

mass $m = 1200\,\text{kg}$

speed $v = 13\,\text{m/s}$

kinetic energy $E_k = ?$

$E_k = \frac{1}{2}mv^2$

$= \frac{1}{2} \times 1200 \times 13^2$

$= 101\,400\,\text{J}$

C ## Conservation of energy

The **principle of conservation of energy** says that energy cannot be created or destroyed. Energy can be transferred but the total amount of energy in existence always stays the same.

Example

C A bungee jumper of mass 70 kg freefalls from a bridge 100 m above a river. His bungee rope breaks his freefall when he is 20 m above the river. Ignoring the effects of air resistance:

a) calculate his kinetic energy just at the end of the freefall

b) calculate his speed just at the end of the freefall.

Solution

Gravitational potential energy at the bridge is all transferred to kinetic energy at the end of the freefall.

C a) height of freefall $h = 100 - 20 = 80\,\text{m}$

mass $m = 70\,\text{kg}$

kinetic energy $E_k = ?$

$E_p = mgh$

$= 70 \times 10 \times 80$

$= 56\,000$

so $E_k = 56\,\text{kJ}$

b) kinetic energy $E_k = 56 \times 10^3\,\text{J}$

mass $m = 70\,\text{kg}$

speed $v = ?$

$E_k = \frac{1}{2}mv^2$

$\therefore 56\,000 = \frac{1}{2} \times 70 \times v^2$

$\therefore v = \sqrt{\dfrac{2 \times 56\,000}{70}}$

$= 40\,\text{m/s}$

> ### What You Should Know
>
> ◆ About work done and energy transferred.
> ◆ How to do calculations using the relationship work done = force × distance.
> ◆ About power.
> ◆ How to do calculations using the relationships power = $\dfrac{\text{work done}}{\text{time}}$ or
> $\dfrac{\text{energy transferred}}{\text{time}}$.
>
> ◆ About change in gravitational potential energy and work done against or by gravity.
> ◆ About kinetic energy.
> ◆ The energy transformations during different stages of a journey.
> ◆ How to do calculations using the relationship $E_k = \frac{1}{2}mv^2$.
> ◆ About the principle of conservation of energy.
> ◆ How to do calculations involving the principle of conservation of energy.

> ### Answer

SAQ 11

a) Pedalling to increase speed – chemical energy to kinetic energy.
b) Freewheeling up a hill – kinetic energy to gravitational potential energy.
c) Freewheeling down a hill – gravitational potential energy to kinetic energy.
d) Braking to slow down – kinetic energy to heat energy.

Chapter 6

ENERGY MATTERS

In this chapter we look at our use of energy. At present our main source of energy is burning fossil fuels. Since the reserves of fossil fuels are finite, and their use causes pollution, we have to consider alternative sources of energy.

One of the major uses for all types of fuel is to generate electricity. In the second and third sections of this chapter, we look at how electricity is generated and transmitted to the end users.

The final section looks at heat in the home, including energy conservation in buildings.

Supply and demand

Sources of energy are either **renewable** or **non-renewable**. The best way to think of the difference is that if a source of energy is a fuel that is consumed (used up) in some way, then it is a non-renewable source of energy. If there is no use of a fuel involved, the source is a renewable source of energy.

Question

SAQ 1   The following are all sources of energy:

coal, gas, hydro, nuclear (uranium fuel), oil, peat, solar (from the Sun), tidal, wave, wind.

a) List all the fossil fuels.
b) List all the non-renewable sources of energy.
c) List all the renewable sources of energy.
d) **Compare your list of fossil fuels with your list of non-renewable sources of energy. What is the odd one out?**

Using renewable energy sources conserves the limited stocks of fossil fuels. They can also be used to produce electricity directly – no fuel is burnt producing pollutants or carbon dioxide. So there is less acid rain and the 'greenhouse effect' in the atmosphere is reduced.

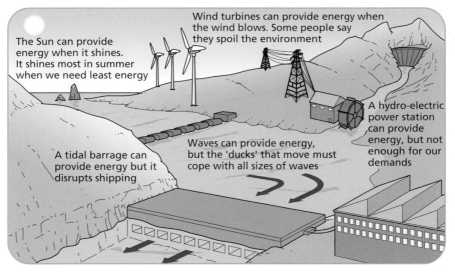

Figure 6.1 Renewable energy sources

The labels in Figure 6.1 read:

The Sun can provide energy when it shines. It shines most in summer when we need least energy

Wind turbines can provide energy when the wind blows. Some people say they spoil the environment

A hydro-electric power station can provide energy, but not enough for our demands

A tidal barrage can provide energy but it disrupts shipping

Waves can provide energy, but the 'ducks' that move must cope with all sizes of waves

Question

SAQ 2 Use Figure 6.1 to explain a disadvantage of any three of the sources shown.

Example

The total power consumption of everyone in the United Kingdom is about 250 GW (gigawatts). Calculate how many power stations, each with an output of 1250 MW, are needed to satisfy this demand.

Solution

total power consumption = 250 GW = $250{\times}10^9$ W
power output per power station = 1250 MW = $1250{\times}10^6$ W

$$\text{number of power stations} = \frac{\text{total power consumption}}{\text{power output per power station}}$$

$$= \frac{250 \times 10^9}{1250 \times 10^6}$$

$$= 200$$

Individually and collectively we use an enormous amount of energy. But there are ways which we can reduce the energy we use by conserving energy. By conserving energy we mean making the most efficient use of energy.

Question

SAQ 3 Figure 6.2 shows ways of conserving energy in the home, in transport and in industry.

Copy and complete the statements to explain how energy is conserved in each case.

Energy is conserved in the home by putting in loft insulation because

----

Energy is conserved in transport by making cars streamlined because

----

Energy is conserved in industry by using more efficient light bulbs because

----

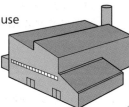

Figure 6.2

What You Should Know

◆ About our use of fossil fuels.
◆ About renewable and non-renewable sources of energy.
◆ How to do calculations about energy supply and demand.
◆ How to conserve energy in the home, in transport and in industry.

Answers

SAQ 1

a) The fossil fuels are coal, gas, oil and peat.
b) The non-renewable sources of energy are coal, gas, nuclear (uranium fuel), oil and peat.
c) The renewable sources of energy are hydro, solar (from the Sun), tidal, wave and wind.
d) Uranium is the only fuel that is not a fossil fuel. Nuclear energy uses a non-renewable source.

Answers *continued*

SAQ 2

Any three from:

◆ The Sun can only provide solar energy when it shines.
◆ Wind turbines can only provide energy when the wind blows.
◆ A hydro-electric power station cannot provide enough energy for our needs.
◆ A tidal barrage disrupts shipping.
◆ Bobbing 'ducks' in waves can supply energy but have to be able to work in all types of waves.

SAQ 3

a) Energy is conserved in the home by putting in loft insulation because this reduces the rate at which heat can escape. This means that less fuel is used in the heating system.
b) Energy is conserved in transport by making cars streamlined because this reduces air resistance. Less work is done by the engine so less fuel is used.
c) Energy is conserved in industry by using more efficient light bulbs because less energy is used to produce the light level needed.

Generating electricity

Power stations

Energy is transferred to consumers as electricity in a power station. There are three types of power station: thermal, hydro-electric and nuclear. All have different sources of energy.

> **Question**
>
> **SAQ 4** What is the source of energy, and how is it released, in:
>   a) a thermal power station
>   b) a hydro-electric power station
>   c) a nuclear power station?

In all types of power station, the energy released from the source is used to drive a turbine that is coupled to a generator which produces electricity.

Hydro-electric power station

Hydro-electric power stations are often used to supply the extra power needed in the early evening when everyone cooks a meal and turns on the television.

At night, when not much electricity is being used, the water in the hydro-electric scheme is pumped up to the top reservoir to be ready for the next period of peak demand. A pumped hydro-electric scheme is an efficient way of generating electricity. It is used to cope with the changing demands for electricity throughout the day and night.

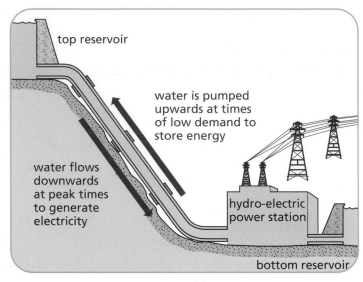

Figure 6.3 A pumped hydro-electric scheme

Example

In a pumped hydro-electric scheme, water is held in a reservoir 40 m above the turbine. The scheme is called on to provide 2 MW of power.

Calculate how much water must flow through the turbine per second to provide this power.

Solution

height $h = 40\,\text{m}$
power $P = 2\,\text{MW} = 2\,000\,000\,\text{W} = 2\,000\,000\,\text{J/s}$

$$E_p = mgh$$
$$2\,000\,000 = m \times 10 \times 40$$
$$\therefore m = \frac{2\,000\,000}{400}$$
$$= 5000\,\text{kg}$$

Nuclear power station

The fuel used in a nuclear power station is uranium. Uranium is not burned in the way that a fossil fuel is burned in a thermal power station. Instead, heat is given off when uranium undergoes **nuclear fission** in a reactor. All you need to know for General Level is that there is radioactive waste produced by nuclear reactors. For Credit Level, try the next two SAQs.

Questions

SAQ 5 Figure 6.4 shows the process of nuclear fission.

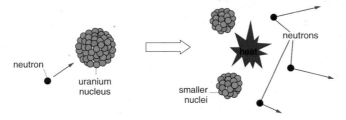

Figure 6.4 Nuclear fission

Use the diagram to help you copy and complete the passage.

When a … hits the … of a … atom, it is absorbed and the nucleus becomes … If there is enough …, the … splits into two smaller … and up to three … are released. Because the smaller … gain … energy, energy in the form of … can be extracted.

SAQ 6 Figure 6.5 shows how a large amount of heat is produced in a chain reaction. Copy and complete the passage below.

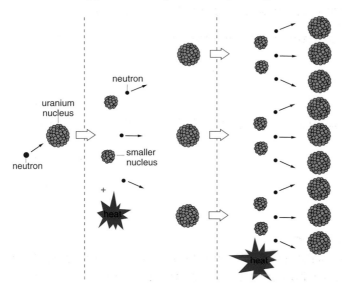

Figure 6.5 A chain reaction

The … of a single … only produces a … amount of heat. However, more … are produced during the fission. These … can go on to cause the … of more … nuclei and each of these fissions produces … and more … This allows further fissions to happen. This process is called a …

Example

C One tonne (1000 kg) of coal can supply $2 \cdot 8 \times 10^{10}$ J of energy.
One gram of uranium can supply $8 \cdot 2 \times 10^{10}$ J of energy.
Compare the energy available per kilogram of uranium and coal.

Solution
energy per kilogram of uranium $= 8 \cdot 2 \times 10^{10} \times 10^{3}$ J $= 8 \cdot 2 \times 10^{13}$ J
energy per kilogram of coal $= 2 \cdot 8 \times 10^{10} \times 10^{-3}$ J $= 2 \cdot 8 \times 10^{7}$ J

$$\therefore \frac{\text{energy per kilogram of uranium}}{\text{energy per kilogram of coal}} = \frac{8 \cdot 2 \times 10^{13}}{2 \cdot 8 \times 10^{7}}$$

$$= 2 \cdot 9 \times 10^{6}$$

C When energy is transferred in a power station, the total amount of energy remains the same (it is conserved) but the amount of useful (electrical) energy is less than the input energy. This is because some energy is lost as heat to the surroundings and cannot be transformed into electrical energy. We say that energy is **degraded** in an energy transformation.

The efficiency of energy transformation is given by the relationship:

$$\text{efficiency} = \frac{\text{useful energy available at output}}{\text{total energy input}} \quad \text{or}$$

$$\text{percentage efficiency} = \frac{\text{useful } E_o}{E_i} \times 100$$

Example

C The energy delivered by the fuel of a power station is 400 MJ every second (400 MW). The useful energy available at the output is 160 MJ every second (160 MW). Calculate the efficiency of the power station.

Solution
input energy $= 400$ MJ
useful energy available $= 160$ MJ

$$\text{efficiency} = \frac{\text{useful energy available at output}}{\text{total energy input}}$$

$$= \frac{160}{400}$$

$$= 0 \cdot 4 \ (= 40\%)$$

What You Should Know

- ◆ About the energy transformations in thermal, hydro-electric and nuclear power stations.
- ◆ About pumped hydro-electric schemes.
- ◆ How to do calculations on energy transformation.
- ◆ About the waste products of nuclear reactors.
- ◆ What a chain reaction is.
- ◆ What is meant by saying that energy is degraded in energy transformation.
- ◆ How to do calculations involving energy output of different fuels.
- ◆ How to do calculations involving efficiency of energy transformation.

Answers

SAQ 4

a) Fossil fuel. The energy is released as heat by burning the fuel.

b) Gravitational potential energy of the water stored in a high reservoir. This energy is released as the water flows down a pipe.

c) Nuclear fuel – uranium. The energy is released as heat during the process of nuclear fission.

SAQ 5

When a neutron hits the nucleus of a uranium atom, it is absorbed and the nucleus becomes unstable.

If there is enough energy, the nucleus splits into two smaller nuclei and up to three neutrons are released.

Because the smaller nuclei gain kinetic energy, energy in the form of heat can be extracted.

SAQ 6

The fission of a single nucleus only produces a small amount of heat. However, more neutrons are produced during the fission.

These neutrons can go on to cause the fission of more uranium nuclei and each of these fissions produces heat and more neutrons. This allows further fissions to happen. This process is called a chain reaction.

HOW TO PASS STANDARD GRADE PHYSICS

Transmitting electricity

Induced voltage

A voltage is **induced** when a wire (conductor) is moved across a magnetic field.

> **Question** (?)
>
> **SAQ 7** Decide whether a voltage will be induced or not in the following cases.
>
>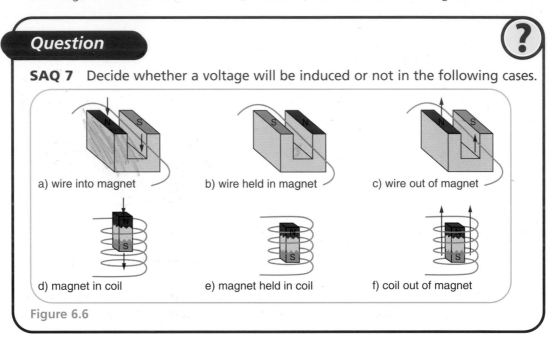
>
> a) wire into magnet   b) wire held in magnet   c) wire out of magnet
>
> d) magnet in coil   e) magnet held in coil   f) coil out of magnet
>
> **Figure 6.6**

C For Credit Level only, you should know that the size of the voltage that is induced depends on the rate at which the conductor cuts through the magnetic field.

> **Question** (?)
>
> **C** **SAQ 8** What are the three factors that could be increased to increase the size of the voltage induced when a magnet moves through a coil?

> **Hints and Tips**
>
> **C** Once again, don't lose marks because of loose language. If you mean to say that the strength of the magnet (or the magnetic field) is increased, don't say 'use a bigger magnet'. 'Big' refers to the size of the magnet, not its magnetic strength. Nowadays there are strong magnets that are very small.

a.c. generators

A **generator** (sometimes called a dynamo) is used to generate a voltage. It does so by moving a coil relative to a magnet as we have just seen. An **a.c. generator** is so called because when it is connected to a circuit, it is an alternating current (a.c.) that is produced. (This might be a useful time to look back at Chapter 2 if you have forgotten what a.c. and d.c. refer to.)

C For Credit Level only, you should be able to explain, using a diagram, how an a.c. generator works.

The coil ABCD is rotated clockwise as shown. Side AB moves up and side CD moves down in the magnetic field. This causes a voltage to be induced in sides AB and CD. These sides act as if they are two batteries connected in series.

When a generator is in a complete circuit, the induced voltage causes a current which lights the lamp. (The current is led to the circuit through the slip rings and the brushes.) As the coil moves past the vertical position (side AB is at the right, CD at the left) the

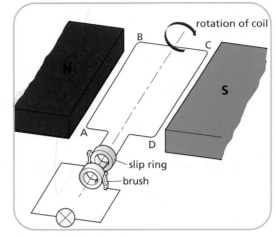

Figure 6.7 A simple a.c. generator

direction of movement in the magnetic field is reversed so the voltage is induced in the opposite sense. This causes the direction of the current to reverse – a.c. is generated.

For Credit Level only, you should be able to state the main differences between a full-size generator and the simple working model that has been described. The main differences concern the magnet and the coil.

◆ Magnet – instead of a stationary magnet, a full-size generator uses rotating electromagnets, called a **rotor**. The rotor is turned by the turbine in a power station. The electromagnets are energised by a small dynamo that turns on the same shaft as the coils.

◆ Coil – instead of a rotating coil, the a.c. voltage is induced in a series of static coils called a **stator**.

Transformers

Transformers change the magnitude of an a.c. voltage. A transformer consists of two coils of wire wound on an iron core. An a.c. voltage is applied to one coil (the **primary coil**) and this induces an a.c. voltage in the other coil (the **secondary coil**).

The relationship between the primary voltage V_p, the secondary voltage V_s, the number of turns on the primary coil n_p and the number of turns on the secondary coil n_s is $\dfrac{V_p}{V_s} = \dfrac{n_p}{n_s}$.

HOW TO PASS STANDARD GRADE PHYSICS

Example

A transformer is designed to change the UK mains voltage of 230 V to that used in the USA, 115 V. There are 2500 turns on the primary coil.

Calculate how many turns are on the secondary coil.

Solution

primary voltage $V_p = 230\,\text{V}$
secondary voltage $V_s = 115\,\text{V}$
number of turns on primary coil $n_p = 2500$
number of turns on secondary coil $n_s = ?$

$$\frac{V_p}{V_s} = \frac{n_p}{n_s}$$

$$\therefore \frac{230}{115} = \frac{2500}{n_s}$$

$$\therefore n_s = 2500 \times \frac{115}{230}$$

$$= 1250 \text{ turns}$$

C For Credit Level only, you should know about the efficiency of a transformer.

The efficiency of a transformer is $\text{efficiency} = \dfrac{\text{power available at secondary}}{\text{power delivered to primary}}$.

Since power is voltage × current, this can be expressed as $\text{efficiency} = \dfrac{V_s I_s}{V_p I_p}$.

Not all the power delivered to the primary of a transformer is available at the secondary. This means that a transformer is not 100% efficient.

Question

SAQ 9 Copy and complete the passage by choosing the correct words.

Because the coils have charge/resistance, some heat/movement is produced in the coils.

Because of the changing electric/magnetic field in the core, small currents/resistances are induced in the core. This produces heat/movement in the core.

Because the core is continually electrified/magnetised and demagnetised, energy is transferred as heat/movement in the core.

Example

C A transformer delivers 540 W at its secondary. The transformer is 90% efficient. The primary voltage of the transformer is 300 V.

Calculate the current in the primary coil of the transformer.

Solution

power available at secondary = 540 W
efficiency = 90%
primary voltage V_p = 300 V
primary current I_p = ?

$$\text{efficiency} = \frac{\text{power available at secondary}}{\text{power delivered to primary}}$$

$$\therefore \frac{90}{100} = \frac{540}{\text{power delivered to primary}}$$

$$\therefore \text{power delivered to primary} = 540 \times \frac{100}{90}$$

$$= 600 \text{ W}$$

$$= V_p I_p$$

$$\therefore I_p = \frac{600}{300}$$

$$= 2 \text{ A}$$

The National Grid

The electricity that is generated in power stations is delivered to houses, shops, factories and all other users by means of a network or grid of transmission lines. These transmission lines, along with a series of transformers, are called the **National Grid**.

The power stations generate electricity at 25 kV. This voltage is increased to 132 kV, 275 kV or 400 kV by **step-up transformers**, and supplied to the National Grid to be transmitted over long distances.

Step-down transformers in sub-stations reduce the voltage to supply all users with electricity at the voltages that they need, varying from 33 kV for heavy industry to 230 V for homes, shops and offices.

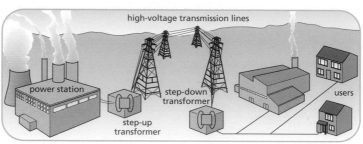

Figure 6.8 The National Grid

Electricity is transmitted at high voltages to reduce power loss. For a certain amount of power transmitted, if the voltage is higher then the current is lower. (Can you remember the relationship between electrical power, current and voltage? If not, look back to Chapter 2.) Heat is produced in the transmission lines because of their resistance. Much more heat would be produced (and so 'lost') if higher currents were used for transmission. High voltages keep the energy loss as low as possible.

Example

Electricity is transmitted along transmission lines that have a resistance of $10\,\Omega$. The electricity is transmitted at a current of $200\,A$.

Calculate the power lost in the transmission lines.

Solution

resistance of lines $R = 10\,\Omega$
current $I = 200\,A$

$$\text{power lost} = I^2R$$
$$= 200^2 \times 10$$
$$= 400\,000\,W\ (= 400\,kW)$$

What You Should Know

- ◆ How a voltage is induced in a conductor.
- ◆ The factors that affect the size of an induced voltage.
- ◆ The main parts of an a.c. generator.
- ◆ How an a.c. generator works.
- ◆ The differences between a full-size generator and a model.
- ◆ About the use and structure of a transformer.

- ◆ How to do calculations using the relationship $\dfrac{V_p}{V_s} = \dfrac{n_p}{n_s}$.

- ◆ Why a transformer is not 100% efficient.
- ◆ How to do calculations involving input and output voltages, primary and secondary currents and efficiency.
- ◆ What the National Grid system is.
- ◆ Why high voltages are used to transmit electricity.
- ◆ How to do calculations involving power loss in transmission lines.

Answers

SAQ 7

a) wire into magnet – voltage is induced
b) wire held in magnet – no voltage is induced (because there is no movement)
c) wire out of magnet – voltage is induced
d) magnet into coil – voltage is induced (there is relative movement between the magnet and the conductor)
e) magnet held in coil – no voltage is induced (again, there is no movement)
f) coil out of magnet – voltage is induced (it doesn't matter whether the coil or the magnet moves)

SAQ 8

The size of the induced voltage is increased by:

◆ increasing the strength of the magnetic field
◆ increasing the number of turns on the coil
◆ increasing the relative speed of movement of the magnet and the coil.

SAQ 9

Because the coils have resistance, some heat is produced in the coils.

Because of the changing magnetic field in the core, small currents are induced in the core. This produces heat in the core.

Because the core is continually magnetised and demagnetised, energy is transferred as heat in the core.

Heat and temperature

It is important that you can explain what is meant by heat and what is meant by temperature.

Questions

SAQ 10 Choose the correct word in each of the following statements and then write out the correct statements in full.

◆ **Heat/Temperature** is a form of energy.

◆ **Heat/Temperature** is a measure of the degree of hotness or coldness of a substance.

◆ **Heat/Temperature** is measured on the Celsius scale.

Questions continued ➢

Questions *continued*

When heat energy is supplied to or taken away from a substance, one of two things will change, but not both at once.

SAQ 11 a) What are the two things that could change when heat is supplied to or taken away from a substance?

b) Can you say which change will happen under different conditions?

The three states of matter are solid, liquid and gas. Adding heat to or taking heat from a substance can change its state.

SAQ 12 Copy and complete the diagram, naming the process in each case.

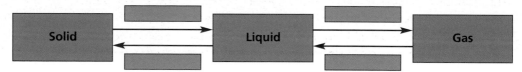

Figure 6.9

Heat can be transferred by three methods:

◆ conduction   ◆ convection   ◆ radiation.

SAQ 13 Match the name of the process with its description by writing out the complete sentences.

a) In conduction …   b) In convection …   c) In radiation …

… the particles of the substance move. This process happens best in liquids and gases.
… no particles are needed. This process takes place in a vacuum.
… the particles vibrate about a fixed position. This process happens best in metals.

Heat travels from an area of high temperature to an area of low temperature. The rate at which heat is lost depends on the temperature difference between the two areas.

SAQ 14 Describe how each of the following reduces heat loss in the home.

a) Double-glazing
b) Cavity wall insulation
c) Loft insulation
d) A jacket round the hot water cylinder
e) Foil-backed plasterboard

Specific heat capacity

Heat and temperature are not the same, but they are related. Adding heat to a substance can increase its temperature. However, the same mass of different substances requires different quantities of heat to raise the temperature by the same amount.

The **specific heat capacity** of a substance, measured in J/kg °C, is the heat needed to increase the temperature of 1 kilogram of the substance by 1 degree Celsius.

The heat E_h needed to change the temperature of m kilograms of a substance of specific heat capacity c by ΔT degrees Celsius is $E_h = cm\Delta T$.

Hints and Tips

The symbol Δ is used to mean 'a change in', so ΔT means 'a change in temperature'. You may have to calculate a change in temperature by finding the difference between an initial temperature and a final temperature.

Example

Calculate how much heat is needed to increase the temperature of 2 kg of water from 20 °C to 90 °C.

Solution

heat $E_h = ?$

mass $m = 2\,\text{kg}$

change in temperature $\Delta T = 90 - 20 = 70\,°\text{C}$

specific heat capacity $c = 4180\,\text{J/kg}\,°\text{C}$ (from data sheet)

$E_h = cm\Delta T$
$= 4180 \times 2 \times 70$
$= 585\,200\,\text{J}$

Hints and Tips

To do a problem like the one above, you need to know the value of the specific heat capacity of water. In the General paper, the value will be given in the question. In the Credit paper, you have to use the value given on the data sheet. Make sure you use the correct table – there are a lot of tables relating to heat on the data sheet.

HOW TO PASS STANDARD GRADE PHYSICS

Example

In the previous example, the heat is supplied to the water by an electric kettle of power 2·5 kW. Calculate how long it takes.

Solution

heat $E_h = 585\,200\,J$
power $P = 2\cdot5\,kW = 2500\,W$
time $t = ?$

$$E_h = Pt$$
$$\therefore t = \frac{E_h}{P}$$
$$= \frac{585\,200}{2500}$$
$$= 234\,s$$

Change of state

We have already seen that heat is involved in a change of state of a substance. This heat is called **latent heat** (latent means hidden – there is no change of temperature when the state changes).

Question

SAQ 15 Copy and complete the sentences below using the words fusion, vaporisation, taken in, given out.

a) When a substance changes state from a solid to a liquid, latent heat of … is …

b) When a substance changes state from a liquid to a solid, latent heat of … is …

c) When a substance changes state from a liquid to a gas, latent heat of … is …

d) When a substance changes state from a gas to a liquid, latent heat of … is …

A refrigerator and a picnic box cooler both make use of a change of state to keep food cool.

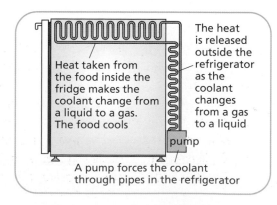

Heat taken from the food inside the fridge makes the coolant change from a liquid to a gas. The food cools

The heat is released outside the refrigerator as the coolant changes from a gas to a liquid

pump

A pump forces the coolant through pipes in the refrigerator

Figure 6.10 The refrigerator

> ### Question
>
> **SAQ 16** Use Figure 6.10 to describe how a refrigerator works.

C The **specific latent heat of fusion** of a substance is the heat needed to change the state of 1 kilogram of the substance at its melting point into a liquid.

> ### Question
>
> **C** **SAQ 17** What is the specific latent heat of vaporisation of a substance?

C The heat E_h needed to change the state of 1 kilogram of a substance at its melting or boiling point is $E_h = ml$ where l is the specific latent heat (of fusion or vaporisation as appropriate) of the substance, measured in joules per kilogram (J/kg).

Example

C Calculate how much heat is needed to convert 0·5 kg of water at its boiling point into steam.

Solution

heat $E_h = ?$ $E_h = ml$

mass $m = 0·5$ kg $= 0·5 \times 2·26 \times 10^6$

specific latent heat of vaporisation $= 1·13 \times 10^6$ J

of water $l = 2·26 \times 10^6$ J/kg (from

data sheet)

> ### What You Should Know
>
>
>
> ◆ About heat and temperature.
> ◆ How to reduce heat loss due to conduction, convection and radiation.
> ◆ What is meant by specific heat capacity.
> ◆ How to do calculations using the relationship $E_h = cm\Delta T$.
> **C** ◆ How to do calculations using the principle of conservation of energy on energy transformations involving a temperature change.
> ◆ What is meant by latent heat of fusion and vaporisation.
> ◆ Applications that involve a change of state.
> **C** ◆ How to do calculations involving specific latent heat.

Answers

SAQ 10

◆ Heat is a form of energy.
◆ Temperature is a measure of the degree of hotness or coldness of a substance.
◆ Temperature is measured on the Celsius scale.

SAQ 11

a) The temperature or the state could change when heat is supplied to or taken away from a substance.
b) When the temperature of the substance is at the melting point or the boiling point, the state will change. Otherwise the temperature will change.

SAQ 12

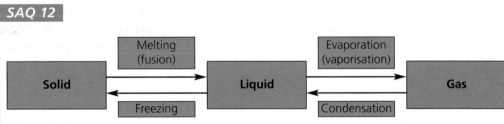

Figure 6.11

SAQ 13

a) In conduction the particles vibrate about a fixed position.
b) In convection the particles of the substance move.
c) In radiation no particles are needed.

SAQ 14

a) Double-glazing traps air (a poor conductor) and so reduces loss by convection also.
b) Cavity wall insulation stops the air in the cavity moving, reducing loss by convection.
c) Loft insulation traps air (a poor conductor) and so reduces loss by convection also.
d) A jacket round the hot water cylinder also traps air and reduces heat loss by convection.
e) Foil-backed plasterboard reflects heat and reduces the rate of heat loss by radiation.

Answers *continued*

SAQ 15

a) When a substance changes state from a solid to a liquid, latent heat of fusion is taken in.
b) When a substance changes state from a liquid to a solid, latent heat of fusion is given out.
c) When a substance changes state from a liquid to a gas, latent heat of vaporisation is taken in.
d) When a substance changes state from a gas to a liquid, latent heat of vaporisation is given out.

SAQ 16

The coolant is pumped around pipes in the refrigerator. When it is in the food compartment, it changes state from a liquid to a gas, taking heat from the food to do so. When the coolant is at the back of the refrigerator it releases this heat as it changes state from a gas to a liquid.

SAQ 17

The specific latent heat of vaporisation of a substance is the heat needed to change the state of 1 kilogram of the substance at its boiling point into a gas.

Chapter 7

SPACE PHYSICS

Well, this is it. The last unit of your Physics revision. The final frontier – space. There are two sections in this chapter. Both introduce some new work, but both also give you the opportunity to go over earlier work.

The first section introduces astronomical terms and also takes earlier work on refraction and lenses a stage further. The spectrum of waves that was introduced earlier is extended and linked together.

The second section starts with some new work on rockets and also looks again at Newton's laws of motion. Finally, the work on satellites at the end of the section covers heat and energy transfer as well as forces and motion.

Signals from space

Astronomy

Remember

You should know the following astronomical terms: **moon**, **planet**, **sun**, **star**, **solar system**, **galaxy**, **universe**, and for Credit Level only, **light-year**.

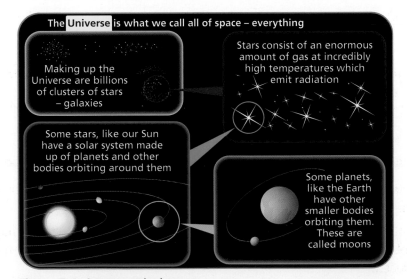

Figure 7.1 Astronomical terms

Question

SAQ 1 Use Figure 7.1 to help you match the following descriptions with these terms: moon, planet, sun, star, solar system, galaxy, universe.

a) A large cluster of stars
b) A body that orbits a planet
c) Part of a solar system orbiting a sun
d) Planets and other bodies that orbit a sun
e) An enormous amount of gas at high temperature emitting radiation
f) A star at a particular stage of its life cycle
g) All of space

You should know approximately how long it takes light to go different distances. This information is useful because astronomical distances are so great.

Questions

SAQ 2 Copy and complete the following by choosing the correct time in each case.

a) It takes light approximately **8 minutes/4 years/100 000 years** to travel from the Sun to Earth.

b) It takes light approximately **8 minutes/4 years/100 000 years** to travel from the next nearest star to Earth.

c) It takes light approximately **8 minutes/4 years/100 000 years** to travel from the edge of our galaxy to Earth.

SAQ 3 A light-year is the distance light travels in 1 year. How many metres are equal to 1 light-year?

The telescope and the magnifying glass

Astronomers sometimes use refracting telescopes to obtain images of space.

A refracting telescope consists of an objective lens and an eyepiece lens in a light-tight tube.

◆ The objective lens is a long focal length convex lens. It produces an image of the object. The larger the diameter of the objective lens, the brighter the image produced.

◆ The eyepiece lens is a short focal length convex lens. It magnifies the image produced by the objective lens.

◆ The length of the light-tight tube can be altered to allow the image to be brought to a focus.

A double star cluster viewed through a refracting telescope

Question

SAQ 4 Use the information given above to draw a diagram showing the main parts of a refracting telescope.

When the object is closer than the focal length from a convex lens (as in the eyepiece of a telescope), the lens acts as a magnifying glass.

Question

SAQ 5 Copy and complete the ray diagram in Figure 7.2 to show how an image is formed by the magnifying glass. If you cannot remember how to

focus   object   focus

Figure 7.2

draw the rays, look back to the 'Light and sight' section of Chapter 3 to remind yourself.

Spectra

White light is made up of different colours. Each colour has a different wavelength. White light can be split into its different colours using a **prism**.

Question

SAQ 6 Use Figure 7.3 to list the colours red, green and blue in order of increasing wavelength.

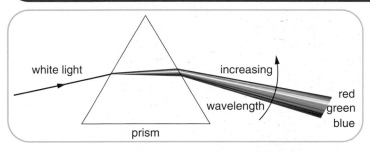

Figure 7.3 **The white light spectrum**

The spectrum of white light produced by a prism is a **continuous spectrum** – all the wavelengths (colours) merge into each other. The spectra produced by some sources, heated elements and stars, for example, are made up of separate lines of colour. These are known as **line spectra**. The line spectrum produced by a source gives information about the atoms in the source.

The electromagnetic spectrum

Light is one member of a large group of waves called the **electromagnetic spectrum**.

◆ Apart from light, the rest of the waves are invisible.
◆ The waves have a wide range of wavelengths (and a wide range of frequencies).
◆ All the waves travel at the speed of light.
◆ The waves or radiations in the electromagnetic spectrum are: gamma rays, X-rays, ultraviolet radiation, visible light, infrared radiation, microwaves, TV waves, radio waves.
◆ The radiations are listed above in order of increasing wavelength and decreasing frequency.

Question

SAQ 7 The following detectors are used for different parts of the electromagnetic spectrum: aerial connected to a tuned circuit; eye; fluorescent materials; Geiger-Müller tube; photoelectric cell; photographic film.

List the waves in the electromagnetic spectrum in order of increasing wavelength, and write the detector or detectors that can be used for each.

We have seen that refracting telescopes detect light from space and that a line spectrum can give information about atoms in a source. Sources also emit other types of radiation in the electromagnetic spectrum. For example, telescopes can detect radio waves. The large telescope at Jodrell Bank in Cheshire is a good example of a radio telescope.

C Other types of radiation can be detected from space. Some sources emit X-rays, ultraviolet radiation and infrared radiation. Different kinds of telescopes are used to detect these signals. These radiations are partly absorbed by our atmosphere. Because of this, telescopes are often sent into space on satellites. The signals they pick up are sent back to Earth. Scientists can get a lot of information about the universe by studying these radiations.

What You Should Know

- ◆ What is meant by the terms moon, planet, sun, star, solar system, galaxy, universe and, for Credit Level only, light-year.
- ◆ The approximate values of some astronomical distances.
- ◆ About the main features of a refracting telescope.
- C ◆ How to draw a ray diagram to show how an image is formed by a magnifying glass.
- ◆ About continuous and line spectra.
- ◆ About the different colours that make up white light.
- ◆ That there is a group of waves that travel at the speed of light.
- C ◆ About the electromagnetic spectrum of radiation – the members in order and detectors for each radiation.
- ◆ Why different kinds of telescope are used to detect signals from space.

Answers

SAQ 1

a) galaxy
b) moon
c) planet
d) solar system
e) star
f) sun
g) universe

SAQ 2

a) 8 minutes
b) 4 years
c) 100 000 years

SAQ 3

1 light-year = number of seconds in 1 year × speed of light in m/s
$$= 365 \times 24 \times 60 \times 60 \times 3 \times 10^8 \text{ metres}$$
$$= 9.5 \times 10^{15} \text{ m}$$

Answers *continued*

SAQ 4

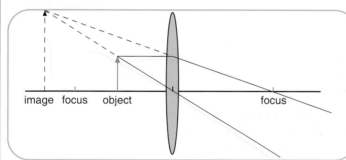

light-tight tube

objective lens

eyepiece lens

Figure 7.4

SAQ 5

image   focus   object   focus

Figure 7.5

SAQ 6

Blue, green, red

SAQ 7

Gamma rays – Geiger-Müller tube, photographic film
X-rays – photographic film
Ultraviolet radiation – fluorescent materials
Visible light – eye, photoelectric cell, photographic film
Infrared radiation – photoelectric cell
Microwaves – aerial connected to a tuned circuit
TV waves – aerial connected to a tuned circuit
Radio waves – aerial connected to a tuned circuit

Space travel

Rockets

A rocket moves because it is pushed forward by propellant gases. In a similar way, the gases are pushed backwards by the rocket.

This is often stated as:

If A pushes on B, then B pushes on A in the opposite direction.

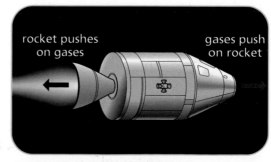

rocket pushes on gases

gases push on rocket

Figure 7.6 Rocket propulsion

HOW TO PASS STANDARD GRADE PHYSICS

C For Credit Level only, you have to take this idea a stage further. You should be able to state **Newton's third law of motion** and identify Newton pairs of forces in different situations.

Remember

C
◆ Newton's third law of motion is 'If A exerts a force on B, then B exerts an equal but opposite force on A'.
◆ The force that A exerts on B and the force that B exerts on A make a Newton pair of forces.

Hints and Tips

C Don't confuse Newton pairs with balanced forces. Balanced forces are equal but opposite forces that act on one object. (Look back to Chapter 5 to re-read the section on balanced forces.) Newton pair forces are equal but opposite forces that act on different objects – A and B.

Question

C **SAQ 8** The forces acting on a boat sailing in the water are shown in Figure 7.7.

Identify the Newton pairs for each of the forces shown.

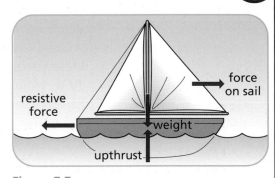

Figure 7.7

You should be able to carry out calculations using thrust, mass and acceleration. If weight is negligible (in outer space, for example) the thrust is the unbalanced force, so

$$\text{thrust} = \text{mass} \times \text{acceleration} \quad F = ma$$

Example

A small rocket attached to an astronaut's backpack supplies 120 N of thrust. The mass of the astronaut is 80 kg. Calculate the acceleration produced.

Solution

thrust $F = 120$ N
mass $m = 80$ kg
acceleration $a = ?$

$$a = \frac{F}{m}$$
$$= \frac{120}{80}$$
$$= 1\cdot5\,\text{m/s}^2$$

A rocket motor is needed to provide an unbalanced force to accelerate a rocket. However, the rocket motor does not need to be kept on once the rocket is moving during interplanetary flight.

Question

SAQ 9 Use one of Newton's laws of motion to explain why a rocket motor need not be kept on during interplanetary flight.

Gravity and weightlessness

There are several facts that you should know about mass, weight, acceleration due to gravity and gravitational field strength. (Some for Credit Level only). You have met some of these before.

Remember

Facts about mass

C
- Mass is a measure of the quantity of matter in an object and is measured in kilograms (kg).
- Any object that has a mass also has a reluctance to have its motion changed. This property is known as **inertia**. An object's mass is a measure of its inertia.

Facts about weight

- Weight is the gravitational pull on an object and is measured in newtons (N).
- Weight is a force.

C
- The weight of an object decreases as the object gets further away from the Earth.
- The weight of an object on the Moon or on different planets is different from the weight of the object on Earth.

Remember continued ➤

Remember *continued*

Facts about acceleration due to gravity and gravitational field strength

◆ The force of gravity near the surface of the Earth gives all falling objects the same acceleration, as long as air resistance can be ignored.

◆ From Newton's second law of motion we get the relationship

$$\text{weight} = \text{mass} \times \text{acceleration due to gravity} \qquad W = mg$$

◆ Gravitational field strength is weight per unit mass.

$$\text{gravitational field strength} = \frac{\text{weight}}{\text{mass}} \qquad g = \frac{W}{m}$$

◆ Acceleration due to gravity and gravitational field strength are different ways of expressing the same quantity.

◆ Objects that fall freely in a gravitational field appear to be weightless. But be careful with this. They *do* have a weight. They only appear to be weightless because they are falling with the same acceleration (the acceleration of gravity) as everything else around them.

Question

SAQ 10 Read through the above facts. Now close the book and write out two facts about each of the quantities mass, weight, acceleration due to gravity and gravitational field strength.

You used the relationship $W = mg$ in the chapter on transport (Chapter 5), but you will also meet situations where g is not equal to 10 m/s^2 (or 10 N/kg).

Example

An astronaut has a mass of 80 kg. Calculate the astronaut's weight on the Moon.

Solution

mass $m = 80$ kg

gravitational field strength on the Moon $g = 1.6$ N/kg (found on data sheet)

weight $W = ?$

$W = mg$
$\quad = 80 \times 1.6$
$\quad = 128$ N

Projectiles and satellites

Projectile motion is a combination of two independent motions:

◆ uniform horizontal speed
◆ uniform vertical acceleration due to the force of gravity.

The combination of these two motions gives a projectile a curved path. The time of flight is the only quantity that is common to both these motions.

Question

C **SAQ 11** A ball is rolled horizontally off a table top as shown in Figure 7.8.

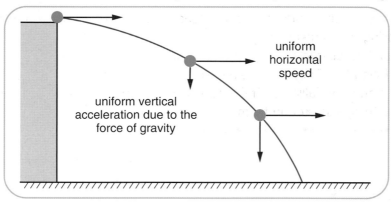

Figure 7.8 Projectile motion

a) Describe and explain the horizontal motion of the ball in terms of Newton's first law of motion.

b) Describe and explain the vertical motion of the ball in terms of Newton's second law of motion.

Example

C A ball is rolled horizontally off a table at 3 m/s as shown in Figure 7.9.

The ball reaches the ground after a time of 0·4 s.

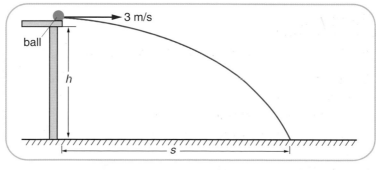

Figure 7.9

Example continued ➤

Example continued

C

a) Calculate the horizontal distance travelled by the ball.
b) Calculate the vertical speed of the ball when it reaches the ground.
c) Draw a graph of the vertical speed of the ball and use it to find the height of the table.

Solution

a) horizontal speed = 3 m/s
time of flight = 0·4 s
horizontal distance = ?

horizontal distance = horizontal speed × flight time
$$= 3 \times 0{\cdot}4$$
$$= 1{\cdot}2 \text{ m}$$

b) initial vertical speed $u = 0$ m/s
acceleration $a = 10$ m/s^2
time of flight $t = 0{\cdot}4$ s
final vertical speed $v = ?$

$$a = \frac{v - u}{t}$$
$$\therefore v = u + at$$
$$= 0 + 10 \times 0{\cdot}4$$
$$= 4 \text{ m/s}$$

c)

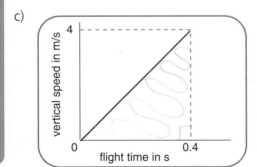

height of table (vertical distance)
= area under graph
$$= \frac{1}{2} \times 0{\cdot}4 \times 4$$
$$= 0{\cdot}8 \text{ m}$$

Figure 7.10

Question

C

SAQ 12 Consider a cannon on the top of a high mountain. Cannonballs are fired horizontally with increasing launch speeds and follow the paths shown at A, B and C in Figure 7.11.

Use this diagram to explain how satellite motion can be considered as an extension of projectile motion.

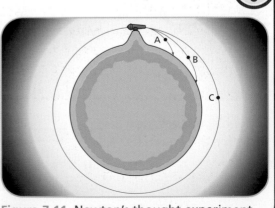

Figure 7.11 Newton's thought experiment

Coming back to Earth

When an object (a spacecraft or a meteor, for example) enters the Earth's atmosphere, a friction force acts on it. This slows it down as work is done against the friction force. The kinetic energy of the object therefore decreases as energy is transferred as heat.

Example

A meteorite consists of a lump of iron of mass 4 kg. On entering the Earth's atmosphere, its speed reduces from 1000 m/s to 500 m/s.

Assuming the energy is all transferred as heat, calculate the increase in temperature of the meteorite.

(The specific heat capacity of iron is 440 J/kg °C.)

Solution

$$\text{initial kinetic energy} = \tfrac{1}{2}\,mv^2$$
$$= \tfrac{1}{2} \times 4 \times 1000^2$$
$$= 2 \times 10^6 \,\text{J}$$

$$\text{final kinetic energy} = \tfrac{1}{2} \times 4 \times 500^2$$
$$= 0\!\cdot\!5 \times 10^6 \,\text{J}$$
$$\therefore \text{loss of kinetic energy} = 1\!\cdot\!5 \times 10^6 \,\text{J}$$

$$E_h = cm\Delta T$$
$$1\!\cdot\!5 \times 10^6 = 440 \times 4 \times \Delta T$$
$$\therefore \Delta T = \frac{1\,500\,000}{440 \times 4}$$
$$= 852\,°\text{C}$$

Hints and Tips

The last exam hint I can give you is 'Stay cool – don't panic!' After all, you have no reason to panic. Since you have reached this far, you have revised all your work thoroughly and are well prepared to sit the exam. Good luck!

What You Should Know

◆ About rocket propulsion.

What you should know continued ➤

What You Should Know *continued*

C
- ◆ About Newton's third law of motion and Newton pairs.
- ◆ How to do calculations using thrust, mass and acceleration.
- ◆ Why a rocket motor need not be kept on during interplanetary flight.
- ◆ Why all falling objects near the Earth's surface have the same acceleration.
- ◆ About the weight of objects on the Moon and on different planets.
- ◆ About objects in free fall.

C
- ◆ What the following terms mean: mass, weight, inertia, gravitational field strength, acceleration due to gravity.
- ◆ How to do calculations using weight, mass, acceleration due to gravity and gravitational field strength.
- ◆ About projectile motion.

C
- ◆ How to do calculations on projectile motion.
- ◆ About satellite motion.
- ◆ About energy transformations during re-entry to the atmosphere.

Answers

SAQ 8

Force on sail: The force of the air (A) on the sail (B) and the force of the sail (B) on the air (A).

Resistive force: The force of the water (A) on the boat (B) and the force of the boat (B) on the water (A).

Upthrust: The force of the water (A) on the boat (B) and the force of the boat (B) on the water (A).

Weight: The pull of the Earth (A) on the boat (B) and the pull of the boat (B) on the Earth (A).

SAQ 9

During interplanetary flight there are no forces to change the motion of the rocket. (There is no friction and there are no gravitational pulls.) Newton's first law of motion tells us that if there are no forces, the rocket will continue at the same speed. So the rocket motor does not need to supply a thrust.

SAQ 10

Any two facts about each of the quantities given in the text.

Answers *continued*

SAQ 11

a) There is no force acting on the ball in the horizontal direction, so by Newton's first law of motion, its horizontal motion is uniform speed. The horizontal distance travelled is given by **horizontal distance = horizontal speed × flight time**. This distance is the area under the horizontal speed–time graph.

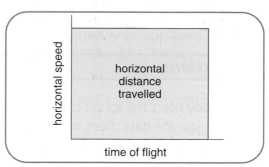

Figure 7.12

b) There is a constant force (the force of gravity or the weight) acting on the ball in the vertical direction. By Newton's second law of motion this means that the vertical motion of the ball is uniform vertical acceleration of $10 \, \text{m/s}^2$. The initial vertical speed is zero, so the vertical speed–time graph is as shown opposite.

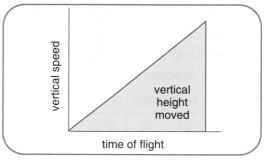

Figure 7.13

The area under this graph gives the vertical height moved by the ball.

SAQ 12

♦ A cannonball fired with a low launch speed would follow path A and hit the Earth near the mountain.

♦ A cannonball fired with a higher launch speed would follow path B and still hit the Earth, but further round.

♦ A cannonball fired with a sufficiently high launch speed would follow path C. This cannonball is continually falling towards Earth with a constant acceleration. Because of the curvature of the Earth, it always misses the Earth as it falls. This cannonball behaves like a satellite in a low-level orbit.

b) What is the function of each of the three parts mentioned?
  i) The amplifier
  ii) The loudspeaker
  iii) The microphone **(3 KU)**
c) The output voltage of the amplifier is 40 V. The resistance of the loudspeaker is 8 Ω.
  i) Calculate the output power of the amplifier. **(2 KU)**
  ii) The input power to the amplifier is 500 mW. Calculate the power gain of the amplifier. **(2 PS)**

9 The following information is taken from a car manufacturer's leaflet about a family car.
  Maximum speed 95·0 miles per hour
  Performance 0 to 27 m/s in 13·5 s
  Mass 1200 kg
  a) Define the term speed. **(1 KU)**
  b) The performance figure gives the maximum acceleration of the car.
    i) What is meant by the term acceleration? **(1 KU)**
    ii) Calculate the maximum acceleration of the car. **(2 PS)**
  c) Calculate the maximum force that the engine of the car can supply. **(2 PS)**

10 A fitness fanatic visits the gym.
  a) He first cycles on an exercise bicycle for 1 minute. During this exercise he pedals 50 turns against a friction force of 40 N. The pedals move a distance of 1·5 m during each turn.
    i) Calculate the work done against friction during this exercise. **(2 KU)**
    ii) Calculate the power developed on the exercise bicycle. **(2 KU)**
  b) The next exercise is pumping weights. During this exercise, a mass of 30 kg is raised 15 times through a distance of 0·5 m.
    i) Calculate how much gravitational potential energy the mass gains during each lift. **(2 PS)**
    ii) Calculate the total work done during this whole pumping exercise. **(1 PS)**

11 In a waterfall, 200 kg of water fall a distance of 20 m every second.
  a) Show that, if all the energy of the water could be used, there is 40 kW of power available from the waterfall. **(3 PS)**
  b) i) Calculate the maximum speed of the water at the bottom of the waterfall. **(2 KU)**
    ii) Explain why the speed of the water at the bottom is less than the maximum calculated in i). **(2 PS)**

12 Every person in the United Kingdom uses energy at a rate of 5000 W. A nuclear power station supplies all the energy needs of all the people in a town. The output of the power station is 1200 MW.
  a) i) State a disadvantage of a nuclear power station. **(1 KU)**

ii) Calculate how many people there are in the town. **(2 KU)**

b) There is a proposal to replace the nuclear power station with wind turbines. The power output of each wind turbine is 3 MW.
   i) State one advantage and one disadvantage of using wind turbines. **(2 PS)**
   ii) Calculate how many wind turbines would be needed to replace the one nuclear power station. **(2 KU)**

13 Heat is constantly being lost in the home.
   a) Describe one way of reducing heat loss in the home due to radiation. **(1 KU)**
   b) On a winter day, the outside temperature is −2 °C. On a summer day, the outside temperature is 18 °C.
   Explain on which day heat is lost more quickly from a house kept at a temperature of 20 °C. **(2 PS)**
   c) A water cylinder in a house holds 200 kg of water.
   Calculate how much energy is needed to increase the temperature of this water from 10 °C to 60 °C. **(3 PS)**

14 A rocket of mass 10 000 kg is at rest on a launch pad on Earth.
   a) i) Calculate the weight of the rocket. **(2 KU)**
   ii) The weight of the rocket is caused by the pull of the Earth on the rocket. This pull forms one half of a Newton pair of forces. State, in words, what the other half of this Newton pair is. **(2 PS)**
   b) The rocket motor supplies an upward thrust of 102 000 N.
   i) Calculate the unbalanced force on the rocket. **(2 KU)**
   ii) Calculate the initial acceleration of the rocket. **(2 KU)**

15 An aid package is dropped from an airplane that is flying horizontally at a speed of 60 m/s. The parcel reaches the ground after 5 s. Air resistance is negligible.
   a) State the horizontal speed of the package when it reaches the ground. **(1 KU)**
   b) Calculate the horizontal distance the package travels. **(2 KU)**
   c) i) Sketch the graph of the vertical speed of the package while it is in the air. **(3 PS)**
   ii) Calculate the height of the airplane when the package was dropped. **(2 KU)**

Answers

1 a) i) speed of sound in air
   = 340 m/s $\frac{1}{2}$
   ii) speed of light in air
   = 3×10^8 m/s $\frac{1}{2}$ (1)

b) $s = vt$ $\frac{1}{2}$
   = 340×3 $\frac{1}{2}$
   = 1020 m **1** **(2)**

c) The speed of light is so fast that it can be taken as instantaneous. **(1)**

2 a)

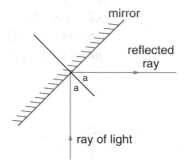

mirror

reflected ray

a
a

ray of light

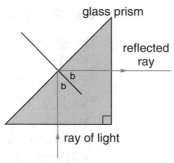

glass prism

reflected ray

b
b

ray of light

2 + 2 **(4)**

b) Either of the following:
   Not all of the light is reflected by a mirror.
   A mirror can give a double reflection. **(1)**

3 a) The tuner **(1)**
   b) i) 3×10^8 m/s **(1)**
      ii) $s = vt$ $\frac{1}{2}$
         = $3 \times 10^8 \times 300 \times 10^{-6}$ $\frac{1}{2}$
         = 90 000 m (= 90 km) **1** **(2)**

c) A colour picture tube contains three electron guns. **1**
   Each gun directs a beam of electrons at one colour of phosphor dots on the screen, red, green or blue. **1** **(2)**

4 a) i) 230 V **(1)**
      ii) $I = P/V$ $\frac{1}{2}$
         = $\dfrac{460}{230}$ $\frac{1}{2}$
         = 2 A **1** **(2)**
      iii) 3 A fuse **(1)**
      iv) $Q = It$ $\frac{1}{2}$
         = $2 \times 5 \times 60$ $\frac{1}{2}$
         = 600 C **1** **(2)**

b) $E = Pt$ $\frac{1}{2}$
   = $460 \times 5 \times 60$ $\frac{1}{2}$
   = 138 000 J **1** **(2)**

c) So that the motor can operate on a.c. **(1)**

5 a)

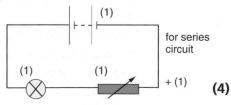

(1)

for series circuit

(1) (1)

+ (1) **(4)**

b) i) $R = V/I$ $\frac{1}{2}$
      = $4/0.4$ $\frac{1}{2}$
      = 10 Ω **1** **(2)**
   ii) voltage across lamp = 8 V **(1)**

HOW TO PASS STANDARD GRADE PHYSICS

6 a) 50 Hz and 100 Hz **(1)**
  b) Some of the energy is absorbed
    by the ear protectors **1**
    so not all of it reaches the
    ears. **1** **(2)**
  c) i) Ultrasound frequencies are
    beyond the range of human
    hearing. **(1)**
    ii) Monitoring an unborn baby
    in the womb. **(1)**

7 a) ionisation **(1)**
  b) The radiation from various
    sources that is always present
    all around. **(1)**
  c) After correcting for background
    radiation, count rate from
    source falls from 500 counts
    per minute to 125 counts per
    minute. **1**
    after one half-life, count rate =
    250 counts per minute
    after two half-lives, count rate
    = 125 counts per minute **1**
    so 24 hours represents
    2 half-lives
    so half-life = 24/2
    = 12 hours **1** **(3)**
  d) becquerel (Bq) **(1)**

8 a)

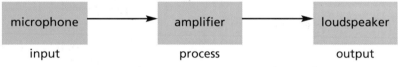

input   process   output

(all correct – 2, 1 or 2 correct – 1)

  b) i) The amplifier boosts/
    increases the strength of
    the input signal.
    ii) The loudspeaker transforms
    the electrical signal into
    sound.
    iii) The microphone transforms
    the sound energy into
    electrical energy. **(3 × 1)**
  c) i) $P = V^2/R$ $\frac{1}{2}$
    $= 40^2/8$ $\frac{1}{2}$
    $= 200$ W **1** **(2)**
    ii) power gain
    = output power/input
    power $\frac{1}{2}$
    $= 200/(500{\times}10^{-3})$ $\frac{1}{2}$
    $= 400$ **1** **(2)**

9 a) Speed is distance divided by
    time. **(1)**
  b) i) Acceleration is the rate of
    change of speed. **(1)**
    ii) $a = \Delta v/t$ $\frac{1}{2}$
    $= 27/13{\cdot}5$ $\frac{1}{2}$
    $= 2$ m/s^2 **1** **(2)**
  c) $F = ma$ $\frac{1}{2}$
    $= 1200 \times 2$ $\frac{1}{2}$
    $= 2400$ N **1** **(2)**

10 a) i) $E_w = Fs$ $\frac{1}{2}$
    $= 40 \times 1{\cdot}5 \times 50$ $\frac{1}{2}$
    $= 3000$ J **1** **(2)**
    ii) $P = E_w/t$ $\frac{1}{2}$
    $= 3000/60$ $\frac{1}{2}$
    $= 50$ W **1** **(2)**

b) i) $E_p = mgh$ $\frac{1}{2}$
$= 30 \times 10 \times 0.5$ $\frac{1}{2}$
$= 150$ J **1** **(2)**

ii) $E_w = 150 \times 15$ $\frac{1}{2}$
$= 2250$ J $\frac{1}{2}$ **(1)**

11 a) In 1 second, E_p lost
$= mgh$ $\frac{1}{2}$
$= 200 \times 10 \times 20$ $\frac{1}{2}$
$= 40\,000$ J **1**
so power available $= 40\,000$ J/s
$= 40$ kW **1** **(3)**

b) i) E_p lost $= E_k$ gained $\frac{1}{2}$
$40\,000 = \frac{1}{2} \times 200 \times v^2$ $\frac{1}{2}$
$v = 20$ m/s **1** **(2)**

ii) Not all the potential energy is transferred as kinetic energy **1**
some is transferred as heat and sound. **1** **(2)**

12 a) i) Radioactive waste is produced by a nuclear power station. **(1)**

ii) number of people
$=$ total power output/
power per person $\frac{1}{2}$
$= 1200 \times 10^6 / 5000$ $\frac{1}{2}$
$= 240\,000$ people **1** **(2)**

b) i) Advantage – does not produce carbon dioxide. **1**
Disadvantage – power output variable because the wind does not always blow. **1** **(2)**

ii) number of wind turbines
$=$ total power needed/
output per turbine $\frac{1}{2}$
$= 1200 \times 10^6 / 3 \times 10^6$ $\frac{1}{2}$
$= 400$ turbines **1** **(2)**

13 a) By using foil-backed plasterboard. **(1)**

b) More heat is lost on the winter day **1**
because the temperature difference is greater. **1** **(2)**

c) $c = 4180$ J/kg °C **1**
$E_h = cm\Delta T$ $\frac{1}{2}$
$= 4180 \times 200 \times 50$ $\frac{1}{2}$
$= 4.18 \times 10^7$ J **1** **(3)**

14 a) i) $W = mg$ $\frac{1}{2}$
$= 10\,000 \times 10$ $\frac{1}{2}$
$= 100\,000$ N **1** **(2)**

ii) The pull **1**
of the rocket on the Earth. **1** **(2)**

b) i) unbalanced force
$=$ upwards thrust –
weight $\frac{1}{2}$
$= 102\,000 - 100\,000$ $\frac{1}{2}$
$= 2000$ N **1** **(2)**

ii) $a = F/m$ $\frac{1}{2}$
$= 2000/10\,000$ $\frac{1}{2}$
$= 0.2$ m/s^2 **1** **(2)**

15 a) horizontal speed is constant at 60 m/s **(1)**

b) $s = vt$ $\frac{1}{2}$
$= 60 \times 5$ $\frac{1}{2}$
$= 300$ m **1** **(2)**

c) i) **(3)**

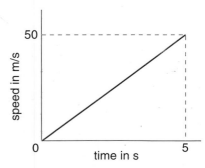

ii) height $=$ area under graph $\frac{1}{2}$
$= \frac{1}{2} \times 5 \times 50$ $\frac{1}{2}$
$= 125$ m **1** **(2)**